I WILL ALWAYS PLACE THE

MISSION F

I WILL

NEVER ACCEPT DEFEAT.

INTRODUCTION TO LEADERSHIP

MSL I

I WILL NEVER QUIT.

I WILL NEVER LEAVE

A FALLEN COMRADE.

Printed in the United States of America

10 9 8 7 6 5 4 3 2 1

ISBN 0-536-97149-8

2005420031

EH/JS

Please visit our web site at *www.pearsoncustom.com*

PEARSON CUSTOM PUBLISHING
75 Arlington Street, Suite 300, Boston, MA 02116
A Pearson Education Company

CONTENTS

Values and Ethics Track

Officership Track

Tactics and Techniques Track

INTRODUCTION

Key Points

Leadership is intangible, and therefore, no weapon ever designed can replace it.

General of the Army Omar N. Bradley

Overview of the BOLC I: ROTC Curriculum

Being an officer in the US Army means being a leader, counselor, coach, strategist, and motivator. Officers must lead other Soldiers in all situations and adjust in environments and situations that are constantly changing. To prepare prospective officers to meet this challenge, the Army ROTC program is designed to develop confident, competent, and adaptive leaders with the basic military science and leadership foundations necessary not only to lead small units in the Contemporary Operating Environment (COE) but also to evolve into the Army's future senior leaders.

The ROTC program is the first, or pre-commissioning, phase of the Army's Basic Officer Leader Course (BOLC). The goal of BOLC is to develop competent and confident leaders in all branches, imbued with the Warrior Ethos, grounded in tactical skills, and skilled in leading Soldiers, training subordinates, and employing and maintaining equipment. BOLC is designed to ensure a tough, standardized, small-unit leadership experience that flows progressively from the pre-commissioning phase (BOLC I, one source of which is ROTC) through the initial-entry field leadership phase (BOLC II) to the branch technical phase (BOLC III). This progressive sequence will produce officers with maturity, confidence, and competence who share a common bond—regardless of whether their branch is in combat arms, combat support, or combat service support—and who are prepared to lead small units upon arrival at their first unit of assignment.

The foundation of the Army ROTC Military Science and Leadership (MSL) Curriculum is the BOLC common core task list, which represents the establishment of competencies a second lieutenant needs to have developed upon arrival at his or her first unit. ROTC cadets receive education and training in the majority of the BOLC tasks, as do officers produced by other commissioning sources (the US Army Military Academy and Officer Candidate School). Then, in BOLC II and III, all second lieutenants, regardless of commissioning source, participate in more-advanced, field- and branch-oriented training exercises in which these tasks are integrated and reinforced.

As with BOLC, ROTC's MSL courses are sequential and progressive. The MSL courses are organized into five topical tracks: Leadership, Personal Development, Values and Ethics, Officership, and Tactics and Techniques. Instruction in these five tracks is sequenced into the Basic and Advanced Courses. The Basic Course (MSL I and II, normally the cadet's freshman and sophomore years) is designed to enhance student interest in ROTC and the Army while providing an overview of each of the five MSL tracks. The Advanced Course (MSL III and IV, plus the Leader Development and Assessment Course (LDAC)) then develops each MSL track in greater depth in order to teach the cadet all the knowledge, skills, and attitudes essential for commissioning, success at BOLC II and III, and the establishment of a sound foundation for a career as a commissioned Army officer.

In addition to classroom instruction, the ROTC program features multiple opportunities for cadets to apply military science and leadership concepts in a variety of environments. Required for all contracted cadets, Leadership Labs meet a minimum of one hour per week to provide cadets with practical experience in applying leadership dimensions first learned in the classroom. For further leadership development and mastery of tactical skills, ROTC battalions also conduct at least two field training exercises (FTXs) per year, each lasting at least 24 hours. Contracted cadets also must participate in physical training (PT) to ensure they acquire a fitness ethos and meet and maintain Army Physical Fitness Test (APFT) standards. The crucible of the ROTC program is cadet attendance of LDAC, normally between the MSL III and IV years. The primary focus at LDAC is to evaluate each cadet's officer potential in a collective garrison and field training environment. The secondary purpose of LDAC is to validate specific skills taught on campus and to teach selective individual and collective common skills.

Military Science and Leadership Tracks

Each of the five learning tracks in the Army ROTC MSL curriculum has sub-sections that are reiterated and developed progressively throughout the MSL courses. The US Army has long recognized the importance of an effective leader who fully embodies the leadership ethos, who is fully committed to being a lifelong learner of leadership as a process and journey rather than destination, and who has the professional acumen to put this leadership into action in an effective, value-added manner regardless of the challenges faced in the fast-paced, ever-changing COE. This is described in Army leadership doctrine as *Be, Know, Do*.

Leadership

- *Leader Attributes* from FM 22–100 are used throughout the curriculum as a graphic organizer for developing a basic knowledge of leader dimensions. Because attributes represent the "*Be*" of Army leaders, the implicit focus throughout the curriculum is on the importance of personal discipline in controlling the mental, physical, and emotional dimensions of life.

- *Leader Skills* (interpersonal, conceptual, technical, and tactical) are defined and illustrated as they apply to direct (tactical), organizational (operational), and strategic levels of leader responsibility. The course of study as a whole is designed to challenge and develop the interpersonal (communication, motivation), conceptual (critical thinking, problem solving), technical (using equipment, predicting effects), and tactical (ordered arrangement and maneuver of combat elements) skills of the leader. This knowledge base represents the "*Know*" of Army leadership.

- *Leadership Actions* teach cadets to become increasingly aware of their own leadership behaviors, including their strengths and weaknesses. These actions represent the "*Do*" of Army leadership.

Personal Development

- *Physical Fitness* is foundational for Army leader development. Every cadet who seeks to become an officer must be able to demonstrate an exceptional level of physical fitness, stamina, and mental toughness.

- *Communication Skills* are essential for success in any leadership arena. For Army officers, however, the effectiveness of a written or oral message can mean the difference between life and death and mission success or failure.

- *Flexibility* has always been and continues to be an imperative characteristic for officers serving in the US Army. Vignettes and case studies from history and the COE are used to challenge cadets to examine asymmetrical and non-linear situations and apply appropriate leadership decision making processes.

- *Adaptability* to negotiate and navigate through situations in which complex factors interact and change constantly is an aspect of personal development that is reinforced in cadets as they progress through their first college experiences during the MSL I year.

Values and Ethics

- *Army Values.* Character development is an implicit aspect of the ROTC curriculum. Cadets are challenged throughout the course of study to recognize and model the seven Army values of loyalty, duty, respect, selfless service, honor, integrity, and personal courage (LDRSHIP). Cadets are expected to demonstrate these values in their daily interactions with others. Values form the foundation for Army leadership.

- *Professional Ethics.* In addition to the Army values, military codes and regulations govern ethical behavior and decision-making. Cadets apply ethical decision making during case studies and historical vignettes.

- *Warrior Ethos* is embedded in case studies and historical vignettes throughout the curriculum. Cadre members discuss the four basic principles of the Warrior Ethos—place the mission first, refuse to accept defeat, never quit, and never leave a fallen comrade—whenever possible. Cadets apply the Warrior Ethos to increasingly complex situations as they progress.

Officership

- *Military Heritage.* Cadre members teach and model military heritage through daily performance and contact, lab exercises, ceremonies, and interpersonal interactions throughout the ROTC curriculum.

- *Military History.* Cadets review vignettes and case studies, which provide opportunities for critical thinking in evaluating matters such as tactics, leadership styles, problem solving and decision making.

- *Management and Administration.* Cadets learn Army programs, policies, and procedures related to areas such as organization, human resources management, administration, training, and facilities in order to support Army operations.

Tactics and Techniques

- *Tactical Operations.* Cadets develop a practical understanding of the basics of map-reading, land navigation, and tactical maneuvering at the individual, team, and squad levels.

The Role of the MSL I Cadet

BE—The MSL I year is the time to master the basics of being an ROTC cadet—Army values, customs and courtesies, physical fitness, and school success or "life skills" such as goal setting and time management. As potential military officers, you will be challenged to study, practice, and evaluate Army leadership and values as you begin your introduction to the Army.

KNOW—To learn the skills required of a quality officer and leader, you must participate actively in learning through critical reflection, inquiry, dialogue, and group interactions. MSL 101 and 102 will teach you the specific leadership values, skills, and actions described in FM 22–100 as they relate to your development as a future Army lieutenant. Baseline instruction in small-unit tactical operations will be the foundation for future, more challenging tactical exercises. Everyone is responsible for contributing to the success of the learning experience.

DO—Extensive small group discussions and exercises are integrated throughout the MSL 101 and 102 courses. Emerging officers are encouraged to work together as a team and with their instructors in modifying assignments, suggesting agendas, and raising questions for discussion. Collaborative learning is enhanced when students apply what they learn in class by describing relevant lessons learned through experiences outside the ROTC classroom.

MSL 101 Course Overview: Leadership and Personal Development

MSL 101 introduces cadets to the personal challenges and competencies that are critical for effective leadership. Cadets learn how the personal development of life skills such as critical thinking, goal setting, time management, physical fitness, and stress management relate to leadership, officership, and the Army profession. The focus is on developing basic knowledge and comprehension of Army leadership dimensions while gaining a big-picture understanding of ROTC, its purpose in the Army, and its advantages for the student. Cadets must meet the following objectives for MSL 101:

Leadership

- Describe the relationship between leader character and competence
- Identify the sixteen dimensions of the Army Leadership Model

Values and Ethics

- Explain the Warrior Ethos
- List and define the seven Army values

Personal Development

- Define standards for the Army Physical Fitness Test (APFT)
- Write short-term and long-term goals to prepare for APFT
- Define the basic elements of time and stress management

Officership

- Explain the importance of being a model citizen as an Army officer
- React to passing colors, National music, and approaching officers

Tactics and Techniques

- Find on-campus locations by reading a campus map

MSL 102 Course Overview: Introduction to Tactical Leadership

MSL 102 overviews leadership fundamentals such as setting direction, problem solving, listening, presenting briefs, providing feedback, and using effective writing skills. Cadets explore dimensions of leadership values, attributes, skills, and actions in the context of practical, hands-on, and interactive exercises. Continued emphasis is placed on recruitment and retention of cadets. Cadre role models and the building of stronger relationships among the cadets through common experience and practical interaction are critical aspects of the MSL 102 experience. Cadets must meet the following objectives for MSL 102:

Leadership

- Distinguish between leadership values, attributes, skills, and actions
- Illustrate leader influencing, operating, and developing actions

Values and Ethics

- Explain how values impact leadership
- Describe the importance of credibility for effective leadership

Personal Development

- Develop personal mission statement and goals
- Explain the basic elements of Army communication

Officership

- Explain the importance of personal development for officership

Tactics and Techniques

- Describe the components of a fire team and squad
- Describe the three individual movement techniques
- Identify symbols and colors on a military map

Academic Approach

The MSL curriculum is designed to focus on the student (cadet), rather than the instructor or the subject matter. Focusing on the cadet requires student-centered objectives and conscious attention to how cadets react to the instruction received. For effective instruction, cadets need the opportunity to try to work with what has been taught. Too often, instruction is limited to the delivery of information, either through reading assignments, lectures, or slide presentations. Active, student-centered learning, in contrast, is founded on the belief that immersive interaction is central to the learning process. Learning occurs during class in the same way it does outside the classroom: through unstructured and structured experiences in which the cadet interacts with cadre, with the instructional material, and with other cadets. Helpful synonyms for ROTC's student-centered approach to learning are experiential learning, direct experience, discovery learning, experience-based learning, and participatory learning. All of these approaches center around five basic steps:

1. Readiness/openness to the experience

2. The experience itself

3. Reflection upon the experience

4. Analysis, theory or additional information to clarify the relationship between theory and actions, with an understanding of lessons learned regarding any needed changes

5. The opportunity to re-experience (practice in new situations/practical exercises).

How to Use This Textbook

The readings in this textbook have been compiled to prepare the cadet to participate actively and productively in MSL classes and labs. The readings are divided into the five MSL curriculum tracks as follows:

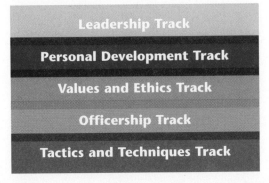

Leadership Track

Personal Development Track

Values and Ethics Track

Officership Track

Tactics and Techniques Track

Color-coded tabs on each page indicate the track. Tracks are then subdivided into sections, which are numbered sequentially within the track.

To be most effective, MSL class sessions are often sequenced to coincide with Leadership Lab schedules, which vary from campus to campus. Thus, class sessions are unlikely to follow the same sequence as textbook sections. Cadets must follow the reading assignments given by their instructors to ensure they are adequately prepared for each class session. The first page of each section orients the cadet to the Key Points to be explained. At the end of each section, learning assessment questions serve as "checks on learning" for the cadet to ensure he or she understands the Key Points of the chapter. Additionally, vignettes, scenarios, and questions are embedded throughout the chapters to help the cadet build critical thinking skills in applying the material read to real-world situations.

Cadet Resources

Cadet Textbook. This textbook contains the readings that support the MSL 101 course, *Leadership and Personal Development*, and the MSL 102 course, *Introduction to Tactical Leadership*.

Cadet CD. The cadet CD (packaged separately) contains additional reference materials, readings, and multimedia that support the MSL program.

Blackboard (*Bb*). The Blackboard course web site, **http://rotc.blackboard.com**, contains MSL course materials.

CONCLUSION

The BOLC core tasks form the foundation of competencies a second lieutenant needs to know upon arrival at his or her first unit. Today's Army officer develops through a progression of BOLC sequential learning programs designed for pre-commissioning (BOLC I), tactical and field training (BOLC II) and branch-specific training (BOLC III). The ROTC program is the implementation of BOLC I in a university setting. Today's ROTC cadet represents the future leadership of our great nation. Such responsibility must be carried by officers well versed in the principles and practices of effective leadership, military operations, and personal development. A future officer and leader must be a confident, competent, and adaptable professional who is ready to lead Soldiers in the COE. The MSL I year of ROTC forges this officer through a challenging curriculum of leader development, Army operations, and personal development. Although this course prepares you for the challenge, it is your responsibility to live the attributes of "Be, Know, and Do" while adopting and demonstrating Army values at all times—both on and off campus. The qualities of an Army officer are not words professed for an exam or exercise. At the MSL I level, these qualities are the expression of a cadet who has learned and looks for opportunities to apply the fundamentals of Army Officership in preparation to "support and defend the Constitution of the United States against all enemies, foreign or domestic." Your commitment to excellence through personal development and diligence in improving your capacity for leadership is essential to the success of the Army of the future.

References

CCR-145-3, *Reserve Officers Training Corps Precommissioning Training and Leadership Development*. 1 September 2005.

Field Manual 22–100, *Army Leadership: Be, Know, Do*. 31 August 1999.

Section

1

INTRODUCTION TO ARMY LEADERSHIP

Key Points

1 What Is Leadership?

2 The *Be, Know, Do* Leadership Philosophy

3 Levels of Army Leadership

4 Leadership vs. Management

5 The Cadet Command Leadership Development Program

Leadership is a potent combination of strategy and character. But if you must be without one, be without the strategy.

GEN Norman Schwarzkopf

Introduction

As a junior officer in the US Army, you must develop and exhibit character—a combination of values and attributes that enables you to see what to do, decide to do it, and influence others to follow. You must be competent in the knowledge and skills required to do your job effectively. And you must take the proper action to accomplish your mission based on what your character tells you is ethically right and appropriate. This philosophy of *Be, Know, Do* forms the foundation of all that will follow in your career as an officer and leader. The *Be, Know, Do* philosophy applies to all Soldiers, no matter what Army branch, rank, background, or gender. SGT Leigh Ann Hester, a National Guard military police officer, proved this recently in Iraq and became the first female Soldier to win the Silver Star since World War II.

Silver Star Leadership

SGT Leigh Ann Hester of the 617th Military Police Company, a National Guard unit out of Richmond, Ky., received the Silver Star, along with two other members of her unit, for their actions during an enemy ambush on their convoy. Hester's squad was shadowing a supply convoy [in March 2005] when anti-Iraqi fighters ambushed the convoy. The squad moved to the side of the road, flanking the insurgents and cutting off their escape route. Hester led her team through the "kill zone" and into a flanking position, where she assaulted a trench line with grenades and M203 grenade-launcher rounds. She and Staff SGT Timothy Nein, her squad leader, then cleared two trenches, at which time she killed three insurgents with her rifle.

When the fight was over, 27 insurgents were dead, six were wounded, and one was captured. Being the first female soldier since World War II to receive the medal is significant to Hester. But, she said, she doesn't dwell on the fact.

"It really doesn't have anything to do with being a female," she said. "It's about the duties I performed that day as a soldier." Hester, who has been in the National Guard since April 2001, said she didn't have time to be scared when the fight started, and she didn't realize the impact of what had happened until much later (Wood, 2005).

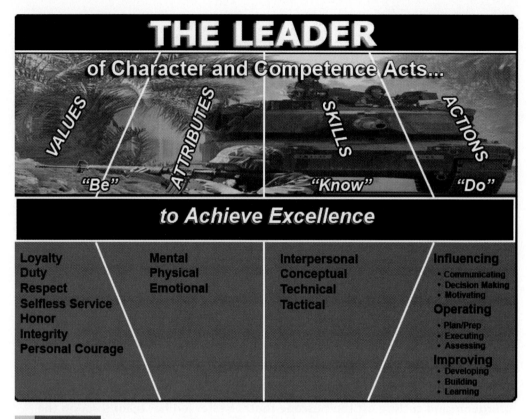

Figure 1.1 The Army Leadership Framework

What Is Leadership?

leadership

influencing people—by providing purpose, direction, and motivation—while operating to accomplish the mission and improving the organization

A simple definition of **leadership** is getting people to do what you want them to do. This *influencing* is a key aspect of the job. But being a leader is a lot more complex than just giving orders. Your influence on others can take many forms. Your words and your deeds, the values you talk about, the example you set, every action you take—on or off duty—are all part of your influence on others. As an Army leader, you will influence your subordinates and others by communicating in three primary ways: *showing purpose*, *giving direction*, and *providing motivation*.

Showing Purpose

By showing purpose, you enable your Soldiers to see the underlying rationale for the mission. As time goes on, your subordinates notice that you communicate in a consistent style of **command** and decision making that builds their trust and confidence. Your Soldiers will eventually be able to "read" a situation and anticipate your intentions and actions. This trust in turn leads to a cohesive, integrated, effective unit.

command

possession and exercise of the authority to command, a specific and legal position unique to the military— the legal and moral responsibilities of commanders exceed those of any other leader of similar position or authority

Giving Direction

When giving direction, you make clear how you want your Soldiers to accomplish the mission. You prioritize tasks, assign responsibility for completing them (delegating appropriate authority), and make sure subordinates understand the Army standard for the tasks. You decide how to accomplish the mission with the available people, time, and resources. It is your subordinates' job to carry out your orders. But to do that, they need clear direction. Give just enough direction to allow Soldiers to use their initiative, abilities, and imagination—and they will surprise you with the results.

Critical Thinking

Discuss the attributes of leadership SGT Hester demonstrated during the ambush in Iraq (see Page 3).

Providing Motivation

Motivation is the will to accomplish a task. By learning about your Soldiers and their capabilities, you will soon be able to gear the team to the mission. Once you have given an order, don't micromanage the process—allow your Soldiers to do their jobs to the best of their abilities. When they succeed, praise them. When they fail, give them credit for the attempt, and coach them on how to improve. Remember that it takes more than just words to motivate. The example you set is at least as important as what you say and how well you manage the work. Stay involved and motivate yourself to attain the best mission result, and your enthusiasm will carry over. Someone once remarked that 90 percent of success is showing up. Your simple presence as a leader can be a powerful morale builder.

The *Be, Know, Do* Leadership Philosophy

The characteristics of an effective Army leader make up the ***Be, Know, Do*** philosophy. As you have already seen, leadership involves influencing others to take appropriate action. But becoming a leader involves much more. Embracing a leadership role involves developing all aspects of yourself: your character, your competence, and your actions. You learn to lead well by adopting Army values, learning military skills, and practicing leadership actions. Only by this self-development will you become a confident and competent leader of character.

> **Be, Know, Do**
>
> *the key characteristics of an Army leader that summarize your a) values and attributes, b) competency and skills, and c) actions and decision making abilities*

BE: Who You Are

Your character is *who you are* and informs everything you do and ask others to do. You demonstrate your commitment to character and to a leadership role in the Army by first adopting and living the seven Army values and attributes (see Tables 1.1 and 1.2). These values form the foundation of your character as a military officer and will guide you in

TABLE 1.1	The Seven Key Army Values
L Loyalty	Bear true faith and allegiance to the US Constitution, the Army, your unit, and other soldiers
D Duty	Fulfill your obligations
R Respect	Treat people as they should be treated
S Selfless Service	Put the welfare of the nation, the Army, and subordinates before your own
H Honor	Live up to all the Army values
I Integrity	Do what is right—legally and morally
P Personal Courage	Face fear, danger, or adversity (physical or moral)

TABLE 1.2	The Three Key Leadership Attributes
Mental	Expressing intellectual capacity and stamina; possessing will, self-discipline, initiative, judgment, and self-confidence—the leader demonstrates strength of mind and the ability to make decisions, even under conditions that strain personal limits
Physical	Projecting the appearance of strength, health, and ability to excel in demanding situations—the leader conveys a professional image of power through military bearing, even under adverse conditions
Emotional	Demonstrating balance, stability, and self-control; maintaining a positive outlook under duress—the leader maintains control through a sense of calm

your career. By living the Army values, you will teach your subordinates by example and help them develop leader attributes.

KNOW: Skills You Have Mastered

Competence in the skills of soldiering is as important as good character in your growth as an Army leader. Without it, your command will lack substance. To ask subordinates to perform to standard, you must first master the standard yourself. You must master four types of Army leadership skills in your training:

- *Interpersonal skills*—knowing your people and how to work with them
- *Conceptual skills*—understanding and applying Army doctrine required to do your job
- *Technical skills*—using your equipment
- *Tactical skills*—making the right decisions in the field

A natural part of an Army officer's career is the opportunity for advancement and promotion. As you advance in rank and responsibility, you will face many new challenges. Having an understanding of and competence in basic Army skills will give you the ability to tackle these new challenges with confidence.

DO: How You Carry out Your Decisions

As you have already seen, leadership takes place in action. What you *Do* is every bit as important as the *Be* and *Know* aspects of your Army leadership philosophy.

While the process of influencing others may seem a little vague or intangible at first, the concept becomes concrete when coupled with *operating actions*. Operating actions are those you take to achieve the short-term goal of accomplishing the mission, such as

TABLE 1.3	Three Leadership Actions
Influencing:	making decisions, communicating those decisions, and motivating people
Operating:	acting to accomplish the unit's immediate mission
Improving:	increasing the unit's ability to accomplish current or future missions

Critical Thinking

Refer to the quote from GEN Schwarzkopf that opened this section. Why would he suggest that *character* takes precedence over strategy? How is what you do informed by who you are?

holding a briefing or conducting a drill. While all direct leaders perform operating actions, the type and scope of such actions become more complex as your rank and level of responsibility change. Moreover, it is a natural part of our competitive drive as humans to want to get better and better at what we do. Leaders, in seeking to build morale, unit **esprit de corps**, and performance, strive to improve the Soldiers, facilities, equipment, training, and resources under their command. Nothing speaks more clearly to your subordinates about your commitment to excellence and improvement than your ongoing assessment of the unit's performance and your leading the way toward improvement. Your investment of time, effort, and interest in your subordinates' improved performance will pay dividends in building trust and *esprit de corps*.

Levels of Army Leadership

Army leadership positions divide into three levels—*direct*, *organizational*, and *strategic*. The leadership level includes a number of factors, including:

- Span of control
- Headquarters level

Figure 1.2 Army Leadership Levels

esprit de corps

a shared sense of comradeship and devotion to the cause among members of a group, team, or unit

Oath of Office Taken by Commissioned Officers in the US Army

I [full name], having been appointed a [rank] in the United States Army, do solemnly swear (or affirm) that I will support and defend the Constitution of the United States against all enemies, foreign and domestic; that I will bear true faith and allegiance to the same; that I take this obligation freely, without any mental reservation or purpose of evasion, and that I will well and faithfully discharge the duties of the office upon which I am about to enter. So help me God.

Leader rank or grade may not automatically indicate the position's leadership level, which is why Figure 1.2 contains no ranks.

- Extent of the influence of the leader holding the position
- Size of the unit or organization
- Type of operations the unit conducts
- Number of people assigned
- The unit's long-term mission or how far in advance it develops plans

Direct Leadership

Direct leadership is face-to-face, first-line leadership. Subordinates of direct leaders see them all the time at the team, squad, section, platoon, company, battery, squadron, and battalion levels. The direct leader may command anywhere from a handful to several hundred people. Direct leaders influence their subordinates one-on-one, but may still guide the organization through subordinate officers and noncommissioned officers (NCOs). Direct leaders quickly see what works, what doesn't work, and how to address problems.

Organizational Leadership

Organizational leaders command several hundred to several thousand people. Their command is indirect, generally through more levels of subordinates. This "chain of command" sometimes makes it difficult for them to see results. Organizational leaders usually employ staffs of subordinate officers to help manage their organizations' resources. Organizational leaders are responsible for establishing policy and the organization's working climate. Their skills are the same as those of direct leaders, but they cope with more complexity, more people, greater uncertainty, and a greater number of unintended consequences. They have little face-to-face contact with the rank-and-file Soldier and command at the brigade through corps levels. Typically, their focus is on planning and missions in the next two to 10 years.

Strategic Leadership

Strategic leaders include military and Department of the Army (DA) civilian leaders from the major command level through the Department of Defense leadership. Strategic leaders are responsible for large organizations and influence several thousand to hundreds of thousands of people. They establish force size and structure, allocate resources, communicate strategic vision, and prepare their commands for their future roles. Strategic leaders consider the total environment in which the Army functions. They may take into account such things as congressional hearings, Army budgetary constraints, new-systems acquisition, civilian programs, research, development, and interservice cooperation.

Leadership vs. Management

As you can see, leadership operates through a wide range of levels, organization sizes, and conditions. Depending on the course of your career as an officer, your path might lead to almost any of these levels and assignments if you are willing to work hard to develop your character, competence, and behavior. You should prepare to embrace the opportunity for promotion when it arises. This path will also take you on an exciting journey through Army life that will almost always provide fulfilling work. One aspect of your job to which you should pay particular attention is the tendency toward the "management mindset." Granted, much of your work as an Army officer will be managerial: putting people and resources to work in the most efficient ways. And managers and good leaders have much in common as both focus on results.

But as Table 1.4 shows, managers and good leaders differ in how they approach their jobs. For example, managers administer, while leaders innovate. And while leaders, like

TABLE 1.4	Management vs. Leadership	
Managers		**Leaders**
Administer		Innovate
Maintain		Develop
Control		Inspire
Short-term view		Long-term view
Imitate		Originate
Ask how/when		Ask what/why
Accept status quo		Challenge status quo

managers, must also keep the organization running smoothly, as a leader you must constantly ponder the next steps, come up with better ways to accomplish the goal, and creatively engage your subordinates to produce more or better results.

In short, leaders continually "push the envelope," searching for ways to change and improve their commands.

Effective leaders build trust and understanding by encouraging their subordinates to seize the initiative and act. They give their Soldiers room to work. This does not mean allowing them to repeat mistakes—your job is to help your subordinates succeed through empowering and coaching. By providing purpose, direction, and motivation for them to operate in support of the mission, you train them to operate independently. A pure management mindset is never able to *let go and lead*.

The Cadet Command Leadership Development Program

The Cadet Command Leadership Development Program is a process designed to develop leadership skills, including those skills you have just reviewed, within a variety of training and educational environments. It is administered on campus by the Professor of Military Science and during summer training by TAC (Train, Advise, Counsel) officers. As you progress through the ROTC program, you will see a variety of different LDP assessment tools that focus on the seven Army values and the 16 leadership dimensions. The Blue Card, the Cadet Evaluation Report, the Officer Evaluation Report, and the Developmental Support Form all share common traits—each drawing on the Army leadership model, which is designed to assist you in maximizing your potential.

You achieve excellence when your Soldiers habitually show discipline and commitment to Army values. Individuals and organizations pursue excellence to improve. The Army

The Art of Delegating Downward

The challenge of command is to empower your subordinate leaders. Give them a task, delegate the necessary authority, and then let them do the work. Check on them frequently enough to keep track of what is going on, but don't get in their way. Your mastery of this skill will improve through practice.

Critical Thinking

How are managers different from good leaders? Can you think of examples of each in your own life? Which would you rather be? Explain.

Just as the diamond requires three properties for its formation—carbon, heat, and pressure—successful leaders require the interaction of three properties—character, knowledge, and application. Like carbon to the diamond, character is the basic quality of the leader. But as carbon alone does not create a diamond, neither can character alone create a leader. The diamond needs heat. Man needs knowledge, study, and preparation. The third property, pressure—acting in conjunction with carbon and heat—forms the diamond. Similarly, one's character, attended by knowledge, blooms through application to produce a leader (FM 22-100).

GEN Edward C. Meyer

cherishes leaders of character who are good role models, consistently set the example, and accomplish the mission while improving their units. The Cadet Command Leadership Development Program is a preview of the Officer Evaluation System, an ongoing performance assessment of regular Army officers, and gives you a foretaste of how others will help you improve your leadership skills.

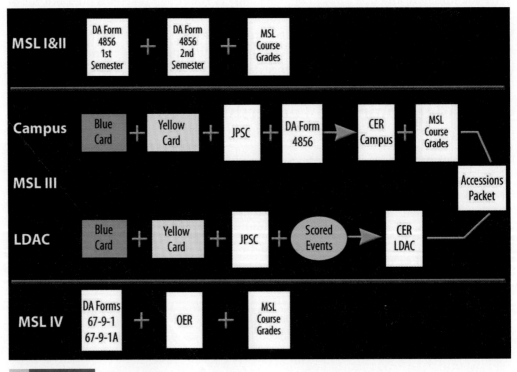

Figure 1.3 *LDP Model*

Cadet Evaluation Report	Type of Report
For use of this form see CC Reg 145-3; staff proponent is USACC DOLD	CAMPUS

PART I - ADMINISTRATIVE DATA

a. NAME (LAST, FIRST, MIDDLE INITIAL)			b. SSN	c. SEX	d. REGION		e. REGT/CO/PLT
f. SCHOOL	g. SCHOOL CODE	h. APFT	i. DATE	j. ESTP (Y/N)	k. HEIGHT		l. WEIGHT

PART II - AUTHENTICATION

(Rated cadet's signature verifies cadet has seen completed part I-VI and the administrative data is correct)

(Rater & Sr. Rater's signatures verify that the cadet has been counseled)

a. NAME OF RATER (LAST, FIRST, MI)	b. SSN	c. RANK	d. POSITION	e. SIGNATURE	f. DATE
g. NAME OF SENIOR RATER (LAST, FIRST, MI)	h. SSN	i. RANK	j. POSITION	k. SIGNATURE	l. DATE
m. RATER'S TELEPHONE NUMBER		n. SENIOR RATER'S TELEPHONE NUMBER			
o. PERIOD COVERED FROM: TO:		p. SIGNATURE OF RATED CADET			q. DATE

PART III - LEADERSHIP POSITIONS

List the evaluated leadership positions from the JPSC (Minimum of 5)

PART IV - PERFORMANCE DATA

NOT USED

PART V - PERFORMANCE EVALUATION - PROFESSIONALISM (Primary Assessor/PLT TAC)

CHARACTER Disposition of the leader: combination of values, attributes, and skills affecting leader actions

a. Values - Indicate "S" or "N" for each OBSERVED value   "N" Ratings must be justified by observation in Part VI below

	S	N		S	N
1. LOYALTY (LO): Bears true faith and allegiance to the Constitution, Army, Units and soldier			5. HONOR (HO): Adheres to ARMY'S CODE OF VALUES		
2. DUTY (DU): Fulfills professional, legal and moral obligations			6. INTEGRITY (IT): Exhibits high personal moral standards		
3. RESPECT (RE): Promotes dignity, consideration, fairness and EO			7. PERSONAL COURAGE (PC): Manifests physical and moral courage		
4. SELFLESS SERVICES (SS): Places Army priorities before self					

b. LEADERSHIP ATTRIBUTES/SKILLS/ACTIONS: Place an "X" in the appropriate rating block for dimension within Attributes, Skills, and Actions.

"E" and "N" must be justified by observations in Part VI.

ATTRIBUTES Fundamental qualities and characteristics	1. MENTAL (ME) Posses desire, will, initiative, and discipline	E	S	N	2. PHYSICAL (PH) Maintains appropriate level of physical fitness and military bearing	E	S	N	3. EMOTIONAL (EM) Display self control; calm under pressure	E	S	N
SKILLS Skill development is a part of self-development; prerequisite to action	4. CONCEPTUAL (CN) Demonstrates sound judgment, critical /creative thinking, moral reasoning	E	S	N	5. INTERPERSONAL (IP) Shows skill with people coaching, teaching, counseling, motivating, and empowering	E	S	N	6. TECHNICAL (TE) Possess the necessary expertise to accomplish all tasks and functions	E	S	N
	7. TACTICAL (TA) Demonstrates proficiency in inquired professional knowledge, judgment, and warfighting	E	S	N								
INFLUENCING Method of reaching goals while operating/improving	8. COMMUNICATING (CO) Display good oral, writing, and listening skills for individuals/ groups	E	S	N	9. DECISION MAKING (DM) Employs sound judgment, logical reasoning, and uses resources wisely	E	S	N	10. MOTIVATING (MO) Inspires, motivates, and guides other toward mission accomplishment	E	S	N
OPERATING Short-term mission accomplishment	11. PLANNING/ PREPARING (PL) Develops detailed executable plans that are feasible, acceptable, and executable	E	S	N	12. EXECUTING (EX) Shows tactical proficiency, meets mission standards, and takes care of people resources	E	S	N	13. ASSESSING (AS) Uses after action and evaluation tools to facilitate consistent improvement	E	S	N
IMPROVING Long-term improvement in the Army; its people and organizations	14. DEVELOPING (DE) Invests adequate time and effort to develop individual subordinates	E	S	N	15. BUILDING (BD) Spends time and resources improving individuals, teams, groups, and units; fosters ethical climate	E	S	N	16. LEARNING (LR) Seeks self-improvement and organizational growth; envisioning, adapting, and leading changes	E	S	N

ROTC CDT CMD FORM 67-9

Figure 1.4 *Sample Cadet Evaluation Report*

NAME	SSN	PERIOD COVERED FROM	TO

PART VI - PERFORMANCE AND POTENTIAL EVALUATION (PLT TAC/Primary Assessor)

a. EVALUATE THE RATED CADET'S PERFORMANCE DURING THE RATING PERIOD AND HIS/HER LEADERSHIP POTENTIAL FOR COMMISSIONING

☐ E-OUTSTANDING PERFORMANCE MUST COMMISSION   ☐ S-SATISFACTORY PERFORMANCE COMMISSION   ☐ N-NEEDS IMPROVEMENT BEFORE COMMISSIONING

b. COMMENT ON SPECIFIC ASPECT OF THE PERFORMANCE AND POTENTIAL FOR COMMISSIONING

c. IDENTIFY ANY UNIQUE PROFESSIONAL SKILLS OR AREAS OF EXPERTISE OF VALUE TO THE ARMY THAT THIS CADET POSSESSES WHICH MAY ASSIST IN DETERMINING BRANCH AND COMPONENT SELECTION

PART VI - SENIOR RATER (PMS)

a. EVALUATE THE RATED CADET LEADER POTENTIAL FOR COMMISSIONING

☐ BEST QUALIFIED   ☐ FULLY QUALIFIED   ☐ QUALIFIED

b. PERFORMANCE COMPARE WITH CADETS IN THE SAME UNIT (Campus Only)

☐ BEST QUALIFIED

☐ FULLY QUALIFIED

☐ QUALIFIED

☐ OTHER

I RANK THIS CADET
_____ OF _____

c. COMMENT ON PERFORMANCE/POTENTIAL

ROTC CDT CMD FORM 67-9

Figure 1.4 *continued*

CONCLUSION

Excellence in leadership does not mean perfection. Rather, an excellent leader allows people room to learn from their mistakes as well as savor their successes. Your subordinates will learn to trust that, when they fail—as they will—you will coach them to do better. As you reflect on the Army values and leadership attributes, developing the skills your position requires, you will become a leader of character and competence. This is the heart of true leadership—influencing people by providing purpose, direction, and motivation while operating to accomplish the mission and improving the organization.

You have learned that effective leadership—particularly in field operations—is your primary and most important challenge as an Army officer. You practice values that lead to excellence and develop a team that can prevail in defense of the United States. The Army expects you, as one of its leaders, to Be, Know, and Do to the very best of your ability. This model is the keystone for high morale and outstanding performance throughout the entire organization.

Learning Assessment

1. What are the components of *Be, Know, Do* in the Army Leadership Model?
2. What are the three levels of Army leadership?
3. Explain the difference between leadership and management attributes.
4. Explain how *Be, Know, Do* applies to your daily activities on campus and in ROTC.
5. What are the key components of the Cadet Command Leadership Development Program at the MSL I and II levels?

Key Words

leadership
command
Be, Know, Do
esprit de corps

References

CCR-145-3, *Reserve Officers Training Corps Precommissioning Training and Leadership Development*, 1 September 2005.

Field Manual 22-100, *Army Leadership: Be, Know, Do*. 31 August 1999.

Schwarzkopf, N. Retrieved 12 July 2005, from http://quotations.about.com/cs/inspirationquotes/a/FamousMilita3.htm

Wood, S. (17 June 2005). Female Soldier Receives Silver Star in Iraq. *American Forces Press Service*. Retrieved 12 July 2005 from http://www4.army.mil/ocpa/soldierstories

ARMY LEADERSHIP— VALUES AND ATTRIBUTES

Key Points

1 What Is Character?

2 Seven Core Army Values

3 Character in Action

4 Three Key Attributes

God grant that men of principle shall be our principal men.

Thomas Jefferson

Introduction

The old wisdom is that an army runs on its stomach. But the physical well-being and ultimate success of an army in the field depends far more on the character of its leaders. Quality leadership grows out of sound values and attributes. Such leadership instills trust, confidence, and loyalty in your subordinates—and produces results. The essential values and attributes of character discussed here will be the moral compass on your journey to becoming a respected, effective leader.

Our values are never tested more strenuously than during times of crisis. Those who can keep a level head and act with character, particularly in the face of grave danger, testify to the importance of the Army's values and attributes framework. Consider how one Army leader in Vietnam reacted with integrity and heroism in a combat situation.

Incident at My Lai

On March 16, 1968, Warrant Officer (WO1) Hugh C. Thompson, Jr., and his two-man crew were on a reconnaissance mission over the village of My Lai, Republic of Vietnam. WO1 Thompson watched in horror as he saw an American Soldier shoot an injured Vietnamese child. Minutes later, when he observed American Soldiers advancing on a number of civilians in a ditch, WO1 Thompson landed his helicopter and questioned a young officer about what was happening on the ground. Told that the ground action was none of his business, WO1 Thompson took off and continued to circle the area. When it became apparent that the American Soldiers were now firing on civilians, WO1 Thompson landed his helicopter between the Soldiers and a group of 10 villagers who were headed for a homemade bomb shelter. He ordered his gunner to train his weapon on the approaching American Soldiers and to fire if necessary. Then he personally coaxed the civilians out of the shelter and airlifted them to safety. WO1 Thompson's radio reports of what was happening were instrumental in bringing about the cease-fire order that saved the lives of more civilians. His willingness to place himself in physical danger in order to do the morally right thing is a sterling example of personal courage (FM 22-100).

Critical Thinking

How did WO1 Thompson exhibit character at My Lai? What was at stake for Thompson? Can you infer anything about the difference between the character of Thompson and that of the young officer he questioned?

What Is Character?

character

who you are, defined by your values, beliefs, and behavior

Character is who you are, demonstrated in what you do. It is the *Be* of the Army's Be-Know-Do values framework. Furthermore, your character defines you as a leader. It informs the decisions you make and the way you treat others. But your character is not something you put on in the morning and take off at night. Rather, it is who you are 24 hours a day, seven days a week, regardless of where you are, whom you are with, or who might be watching.

In your career as an Army officer, you must develop and exhibit character, that combination of values and attributes that enables you to see what needs to be done, decide to do it, and influence others to follow.

Seven Core Army Values

values

the central ideas that form the foundation of your character and guide your decision making and behavior

You demonstrate your commitment to character and to a leadership role in the Army by adopting and living the core Army **values** and attributes. These values form the foundation of your character as a military officer and will guide you in your career. Living the Army values teaches your subordinates by example and helps them develop as individuals and in their own careers.

The core Army values are American values. They grow out of our history and the nature of our society and form the bedrock of the Army's leadership philosophy. By committing them to memory and practicing them, you take part in a time-honored Army tradition of striving for excellent service and ethical behavior.

*The first letters of each of the core Army values form the acronym **LDRSHIP**.*

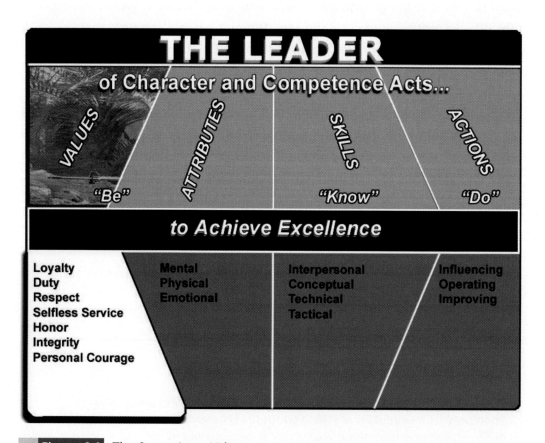

Figure 2.1 The Seven Army Values.

Loyalty

Bear true faith and allegiance to the US Constitution, the Army, your unit, and other Soldiers. Army leaders are loyal to their civilian political leaders. Loyalty ensures the success of the chain of command. As an Army officer, you will find that loyalty is a two-way street. You must be loyal to earn loyalty. Loyalty generates unit cohesion and support among subordinates. No loyalty is fiercer than that of Soldiers who trust their leader to take them through combat. All members of the Army team—active and reserve, National Guard or Army civilians—are loyal to one another.

Duty

Fulfill your obligations. Duty begins with what the law, regulations, and orders require of you, but it is much more than that. Your personal initiative guides you to exceed minimum standards. Army leaders with a strong sense of duty demonstrate and enforce high professional standards at all times. They take the initiative, doing what needs to be done before someone tells them to do it. They take responsibility for their actions and those of their Soldiers.

Respect

Treat people as they should be treated. A very simple way to adhere to this adage is to treat people as you expect to be treated. You demonstrate the value of dignity and human worth, creating a positive command climate and promoting ethnic, racial, religious, and cultural acceptance. You create an environment in which subordinates are challenged and can reach their full potential. The respect you help foster is essential to developing disciplined, cohesive, and effective Army platoons and squads.

Selfless Service

Put the welfare of the nation, the Army, and subordinates before your own. As a member of the Army, you serve the United States. The needs of the Army and the nation come first. You are willing to forgo personal comforts for the sake of others, with no prospect of reward. That doesn't mean, however, that you neglect your family or yourself, as that does the Army more harm than good. Nor does selfless service mean that you can't have a strong ego, high self-esteem, or even healthy ambition. Instead, it means that you don't make decisions or take actions that help your own image or your career but hurt others or sabotage the mission. Team members give of themselves so the team can succeed.

Honor

Live up to all the Army values. Honor provides the moral compass for character and personal conduct in the Army. You demonstrate a keen sense of ethical conduct. You do the right thing, even when no one is looking. Honorable leaders protect the reputation of the profession through their personal actions. For you as an Army leader, honor means putting Army values above self-interest, above career and comfort. Honor is essential for creating a bond of trust among members of the Army and between the Army and the nation it serves. The military's highest award is the Medal of Honor. Its recipients didn't do just what was required of them—they went beyond the expected, above and beyond the call of duty. Some gave their own lives so that others could live. It's fitting that the word we use to describe their achievements is "honor."

Integrity

Do what's right—legally and morally. You are truthful and behave ethically at all times. You act according to accepted principles, not just what might work at the moment. You do the right thing because your character won't permit you to act otherwise. Army leaders

> The essence of duty is acting in the absence of orders or direction from others, based on an inner sense of what is morally and professionally right. . .
>
> GEN John A. Wickham Jr.

say what they mean and mean what they say. You separate right from wrong; you act according to what you know to be right, despite the personal cost; and you say openly that you're acting on your understanding of right and wrong.

Personal Courage

Face fear, danger, or adversity—physical or moral. While fears are a necessary and natural part of human behavior, you are able to weigh the potential cost against the greater need, put fear aside, and do what is necessary to complete your mission. Physical courage means overcoming your fear of bodily harm. Moral courage means standing on your values and principles, even when someone threatens you. Situations involving physical courage are infrequent when you are not in a combat zone; moral courage is required almost daily.

Character in Action

Remember, your acts demonstrate your true character, as you saw with WO1 Thompson at My Lai. Consider how these Soldiers in Operation Desert Storm demonstrated their character and commitment to the Army values, as reported by a platoon sergeant.

Character and Prisoners

The morning of [28 February 1991], about a half-hour prior to the cease-fire, we had a T-55 tank in front of us and we were getting ready [to engage it with a TOW missile]. We had the TOW up and we were tracking him and my wingman saw him just stop and a head pop up out of it. And Neil started calling me saying, "Don't shoot, don't shoot, I think they're getting off the tank." And they did. Three of them jumped off the tank and ran around a sand dune. I told my wingman, "I'll cover the tank, you go on down and check around the back side and see what's down there." He went down there and found about 150 PWs [prisoners of war]. . .

[T]he only way we could handle that many was just to line them up and run them through. . .a little gauntlet. . . [W]e had to check them for weapons and stuff and we lined them up and called for the PW handlers to pick them up. It was just amazing.

We had to blow the tank up. My instructions were to destroy the tank, so I told them to go ahead and move it around the back side of the berm a little bit to safeguard us, so we wouldn't catch any shrapnel or ammunition coming off. When the tank blew up, these guys started yelling and screaming at my Soldiers, "Don't shoot us, don't shoot us," and one of my Soldiers said, "Hey, we're from America; we don't shoot our prisoners." That sort of stuck with me (FM 22-100).

Critical Thinking

Discuss whether character may be especially essential in the face of an enemy who exhibits none. What examples from history or current events can you cite to illustrate this concept?

Three Key Attributes

Attributes are a person's fundamental qualities and characteristics. The three key leader attributes are an individual's mental, physical, and emotional aspects. You can learn and change these attributes. How your individual attributes affect your leadership effectiveness may vary from one situation to the next or from one period in your career to another. The Army expects you to strive to improve your key attributes and to exercise focus, determination, and character as situations dictate.

attributes

the mental, physical, and emotional aspects of the individual

Mental

This refers to your intellectual capacity and stamina. Mental attributes include will, self-discipline, initiative, judgment, self-confidence, and cultural awareness. You express the will to keep going when things are tough and it would be easier to quit. You demonstrate strength of mind and the ability to make decisions, even under conditions that strain your personal limits. You take the initiative and encourage initiative in your Soldiers. You use good judgment to make the best decision possible under the circumstances. You are sensitive to your Soldiers' different backgrounds and to the culture of the country you are operating in. Your mental attributes will be challenged during your ROTC training and college course work.

Physical

Your physical attributes include strength, health, and ability to perform in demanding situations. You work to stay healthy and physically fit. You look like a Soldier—you convey a professional image of power through your military bearing, even under adverse conditions. Your physical attributes will be tested during physical fitness training—and are crucial in combat.

What is life without honor? Degradation is worse than death.

LTG Thomas J. "Stonewall" Jackson

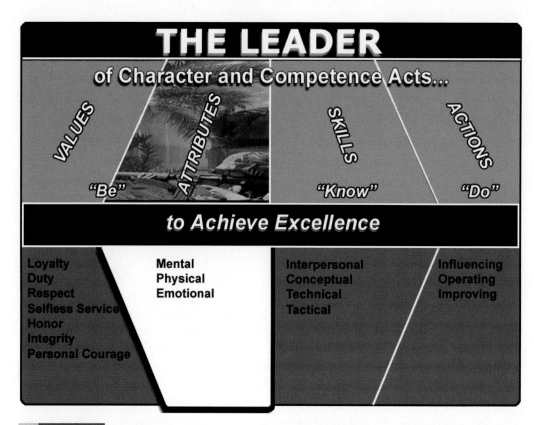

Figure 2.2 Leader Attributes

Emotional

Your emotional attributes include balance, stability, and self-control. You maintain a positive outlook and a sense of calm, even under duress. You understand that emotional energy can bolster or sap will and endurance, giving you a powerful leadership tool. Your balance, stability, and self-control help you make the right ethical choices. You are aware of your own strengths and weaknesses. You are steady under pressure and calm when danger appears. You behave the way you want your Soldiers to behave. Your emotional attributes play a particularly important role in situations of personal change, such as a death, birth, or divorce in your family, breaking up with a girlfriend or boyfriend, or other major life changes.

CONCLUSION

True leaders are made, not born. And true authority derives first from self-control. Soldiers will take direction from officers who themselves appear self-directed. In learning and applying the seven core Army values and working to improve your own key attributes, you will discover that success and respect are natural outcomes of your pursuit of excellence as an Army leader.

Learning Assessment

1. Name the seven core Army values and give an example of how each forms part of the foundation of an effective leader's character. How would the lack of that particular value negatively affect a leader's performance?

2. Consider the three key leadership attributes discussed in this chapter. Would one attribute be more important than the others at certain times in your career?

3. Discuss the difference between values and attributes and give an example of each in your own life.

4. Why are these values and attributes important for an Army leader?

Key Words

character
values
attributes

Reference

CCR-145-3, *Reserve Officers Training Corps Precommissioning Training and Leadership Development*. 1 September 2005.

DA PAM 600-65, *Leadership Statements and Quotes*. 1 November 1985.

Field Manual 22-100. *Army Leadership: Be, Know, Do*. 31 August 1999.

Section

3

ARMY LEADERSHIP— SKILLS

Key Points

1 The Four Army Leadership Skills

2 Acquiring Skills

3 Combining Skills

Use your people by allowing everyone to do his job. When a subordinate is free to do his job, he perceives this trust and confidence from his superiors and takes more pride in his job, himself, and the organization's goals and objectives. Delegation of sufficient authority and proper use of subordinates helps develop future leaders. This is a moral responsibility of every commander (DA PAM 600-65).

LTC Stanley Bonta, in a 1952 Lecture at West Point

Introduction

The Army invests a tremendous amount of time, energy, and resources in training you, its junior leaders, because you will set and maintain standards of excellence in the future.

Direct leadership is the first level of Army leadership. It is face-to-face, first-line interaction. This section explores the skills a direct, first-line leader must master and develop. You will learn more about the Know of Be, Know, and Do for Army leaders. The four essential skill groups introduced are interpersonal, conceptual, technical, and tactical.

In Army units where subordinates see their leaders all the time—teams, squads, sections, platoons, companies, and battalions—direct leaders must master and teach an array of interpersonal, conceptual, technical, and tactical skills. Sometimes teaching essential skills requires making a hard or unpopular choice, as one sergeant discovered at a post in Vietnam.

Rusty Rifles

While serving in the Republic of Vietnam, SFC Jackson was transferred from platoon sergeant of one platoon to platoon leader of another platoon in the same company. SFC Jackson quickly sized up the existing standards in the platoon. He wasn't pleased. One problem was that his Soldiers were not keeping their weapons cleaned properly: rifles were dirty and rusty. He put out the word: weapons would be cleaned to standard each day, each squad leader would inspect each day, and he would inspect a sample of the weapons each day. He gave this order three days before the platoon was to go to the division rest and recuperation (R&R) area on the South China Sea.

The next day SFC Jackson checked several weapons in each squad. Most weapons were still unacceptable. He called the squad leaders together and explained the policy and his reasons for implementing it. SFC Jackson checked again the following day and still found dirty and rusty weapons. He decided there were two causes for the problem. First, the squad leaders were not doing their jobs. Second, the squad leaders and troops were bucking him—testing him to see who would really make the rules in the platoon. He sensed that, because he was new, they resisted his leadership. He knew he had a serious discipline problem he had to handle correctly. He called the squad leaders together again. Once again, he explained his standards clearly. He then said, "Tomorrow we are due to go on R&R for three days and I'll be inspecting rifles. We won't go on R&R until each weapon in this platoon meets the standard."

The next morning SFC Jackson inspected and found that most weapons in each squad were still below standard. He called the squad leaders together. With a determined look and a firm voice, he told them he would hold a formal in-ranks inspection at 1300 hours, even though the platoon was scheduled to board helicopters for R&R then. If every weapon didn't meet the standard, he would conduct another in-ranks inspection for squad leaders and troops with substandard weapons. He would continue inspections until all weapons met the standard. At 1300 hours the platoon formed up, surly and angry with the new platoon leader, who was taking their hard-earned R&R time. The Soldiers could hardly believe it, but his message was starting to sink in. This leader meant what he said. This time all weapons met the standard (FM 22-100).

The Four Army Leadership Skills

Basic military skills form the foundation of the Know aspect of the Army's Be, Know, Do leadership philosophy. To be an effective leader, you must be competent in your field. But **competence** does not necessarily equal excellence; you do not have to be your rifle platoon's best marksman, for example, or the company's fastest runner. Rather, competence means being able to perform to Army standards and by example point the way to improvement, which leads to excellence. You accomplish this goal by gaining knowledge and mastering appropriate skills yourself before you ask others to follow your example.

Interpersonal Skills

Since leadership is about motivating people, it's no surprise to find interpersonal skills—what some call "people skills"—at the top of the Army's list of what a leader must know. Interpersonal skills help you work productively with people. They affect how well you deal with people and include communicating, supervising, and counseling.

competence

mastery of the basic military knowledge and skills necessary to provide leadership

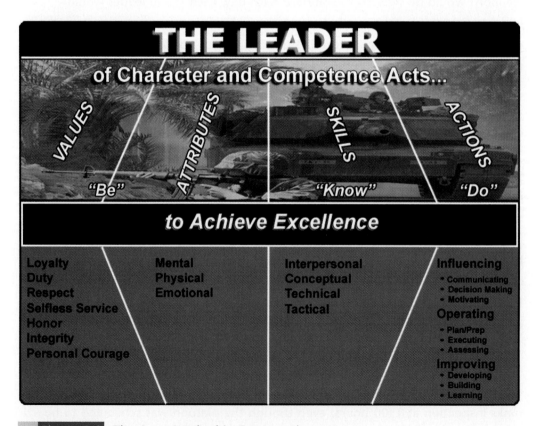

Figure 3.1 The Army Leadership Framework

Critical Thinking

In SFC Jackson's decision to enforce minimal basic weapons standards in his platoon, what Army skills was he attempting to emphasize over his Soldiers' comfort and convenience? Why was such a stand necessary, despite its unpopularity?

Communicating

Since leadership is about getting other people to do what you want them to do, giving them information in a way they clearly understand is a crucial, primary skill. Other interpersonal skills—supervising and counseling for example—depend on your ability to communicate. Communication falls into four broad categories: speaking, reading, writing, and, most important, listening.

As an Army leader, you must train yourself to listen and understand first before you decide what to say. When you practice active listening, you send signals to the speaker that say, "I'm paying attention." And remember: In face-to-face communication, a great deal goes on that is unspoken. Nonverbal communication involves all the signals you send with your facial expressions, tone of voice, and body language. Effective leaders know that communication includes both verbal and nonverbal cues.

Supervising

Good leaders habitually check and recheck things. They find the balance between checking too much and not enough. Train your subordinates to act independently. Give instructions or your intent and then allow your subordinates to work without constantly looking over their shoulders. But don't hesitate to check on aspects critical to the mission (fuel levels), details a Soldier might forget (spare batteries for night vision goggles), or tasks at the limit of what a Soldier has accomplished before (preparing a new version of a report). Checking minimizes oversights, mistakes, or other circumstances that might affect mission success. Checking also gives leaders a chance to see and recognize subordinates who are doing things right or make on-the-spot corrections when necessary. Your role is to answer questions and resolve issues that your squad leaders can't handle.

Counseling

One of the most important aspects of your job as a small unit leader is to develop your subordinates. Effective counseling is the tool you use to help your subordinates develop personally and professionally. As with everything else you do, you must develop your own skills as a counselor. The appropriate counseling you provide will lead to a specific plan of action that your subordinate can use as a road map for improvement. And once the plan of action is set, you must follow up with regular, one-on-one sessions to ensure your subordinate stays on track.

Never tell people how to do things. Tell them what to do and they will surprise you with their ingenuity.

GEN George S. Patton

Conceptual Skills

Conceptual skills enable you to handle ideas productively. Being a sound conceptual thinker requires critical and ethical reasoning, creative thinking, and reflective thinking. Using conceptual skills enables you to analyze processes and problems, predict outcomes, weigh courses of action, create solutions, and make sound decisions.

Critical Reasoning

Critical reasoning is the key to understanding situations, finding causes, arriving at justifiable conclusions, making good judgments, and learning from experience—in short, solving problems. Here "critical" does not mean finding fault; it means thinking about the problem in depth from several points of view instead of just being satisfied with the first answer that comes to mind. Army leaders need this ability because many of the problems they face offer no easy solutions. Sometimes just figuring out what the real problem is presents a huge hurdle. And sometimes you know what the problem is but have no clear answer. At other times, you can come up with two or three workable answers. Your critical reasoning can help you decide which is best.

Ethical Reasoning

Your subordinates count on you to do more than make tactically sound decisions. They rely on you to make decisions that are ethically sound as well. So to be an effective Army leader, you must be able to reason ethically. The right action in the situation you face may not be in regulations or field manuals, because even the most exhaustive regulations can't predict every situation. One of the most difficult tasks you face is determining when a rule simply doesn't apply because the situation falls outside conditions envisioned by the regulation's authors. Ethical leaders do the right things for the right reasons all the time, even when no one is watching. But figuring out what is the "right" thing is often difficult. Your decisions should derive from the Army values you have studied, the institutional culture, and the organizational climate. First, determine what is right by law and regulation. In gray areas requiring interpretation, apply Army values to the situation. Determine the best possible answer from among possible solutions, make your decision, and act on it. Use your knowledge and your experience and be prepared to accept the consequences of your actions. Study, reflection, and ethical reasoning can help. Ethical reasoning is an art, not a science, and sometimes the best answer is hard to determine. In those cases, you must rely on your judgment.

Creative Thinking

All good leaders think creatively. Sometimes you run into a problem that you haven't seen before or an old problem that requires a new solution. Army leaders prevent complacency by finding ways to challenge subordinates with new approaches and ideas. In these cases, rely on your intuition, experience, and knowledge. Ask for input from your subordinates. Reinforce team building by making everybody responsible for difficult tasks. Encourage the use of "outside-the-box" thinking to solve problems.

Reflective Thinking

Good leaders are always striving to become better leaders. This means you consistently assess your strengths and weaknesses and reflect on how to sustain your strengths and correct your weaknesses. To become a better leader, you must also be willing to change. You must be open to feedback on your performance from multiple perspectives—your senior officers, your peers, and your subordinates. You must listen to and use the feedback, which involves reflection. Reflection is the ability to take information, assess it, and apply it to behavior. Everyone makes mistakes, but it's important to learn from those mistakes. Reflection is the means to that end.

Technical Skills

Technical skills are those job-related abilities, including basic Soldier skills such as land navigation, battle drills, and marksmanship. They involve knowing your equipment and its operation.

Knowing Equipment

Sergeants, junior officers, and warrant officers are the Army's technical experts and best teachers. Technical skill means facility with things—equipment, weapons, systems—everything from the towing winch on the front of a vehicle to the computer that keeps track of corps personnel. Direct leaders know their equipment and how to operate it. You are responsible for solving problems with equipment, for figuring out how to make it work better, how to apply it, how to fix it—even how to modify it. Subordinates expect you to know their equipment and be an expert in all the applicable technical skills.

Operating Equipment

As a direct leader, you will know how to operate your equipment and make sure your Soldiers do, as well. You set the example with a hands-on approach. When new equipment arrives, find out how it works, learn how to use it yourself, and train your subordinates to use it. Your Soldiers will tend to respect you more if they see that you are willing to get involved in every aspect of working with new equipment—from maintaining that equipment to mastering its application in training and war. The result is a team that is more than just a collection of trained individuals—it's an organization that is capable of working together.

Tactical Skills

Tactics is the art of employing resources to win battles. Tactical skills employ the other leadership skills in a coordinated fashion. They include squad and platoon tactics, offensive and defensive tactics, and reconnaissance. As part of mastering tactics, you need to ensure that your Soldiers have mastered **fieldcraft**—those skills they need to sustain themselves in the field. Proficiency in fieldcraft reduces the likelihood Soldiers will become casualties. Army leaders gain proficiency in fieldcraft through training, study, and practice. During peacetime, it's up to you to enforce tactical discipline, to make sure your Soldiers practice the fieldcraft skills that will keep them from becoming casualties—in battle or out of it—during operations.

tactics

the art and science of employing available means to win battles and engagements

fieldcraft

the skills a Soldier needs to live in the field, such as practicing daily hygiene, building a shelter, or employing survival techniques

Acquiring Skills

Units fight the way they train. As a direct leader in an Army small unit, you are your Soldiers' primary tactical trainer, both for individuals and for teams. The best way to improve individual and collective skills is to train under simulated combat conditions. Distractions will always pull you away from training your unit. And while it may not always be possible for your unit to train together, planning and scheduling small unit training is a critical part of maintaining unit readiness and morale.

The Army provides you, as the small unit leader, a number of resources to help your Soldiers acquire the necessary skills. One of your most important resources is the built-in expertise of your NCOs, warrant officers, and Department of the Army civilians and contractors. In addition, making use of resources at the company and division level is a way to leverage resources and multiply your unit's training success. That's what it means to say a leader is resourceful—the leader finds and exploits ways to improve individual and unit skills.

Critical Thinking

Explain why making mistakes is an essential part of acquiring a skill. Describe a skill you mastered only after several attempts.

A good leader continually seeks such opportunities for skills training and emphasizes mastery at the individual and unit levels. People learn best when they have incentive to improve. Motivating the mastery of skills can take many forms and might include such incentives as encouraging inter-squad competition, writing letters of commendation, and awarding decorations for skills excellence.

Combining Skills

The four leadership skills are interdependent and support one another. If you know how to aim and shoot a rifle (technical), but cannot communicate with the Soldiers around you (interpersonal) or determine where to deploy troops for the best firepower (tactical) or analyze the strength of the enemy or the best use of the terrain (conceptual), you will not be able to lead in combat. Mastery of all four leadership areas will enable you to demonstrate your competence as a leader and serve as an effective model for members of your command.

Listening Is Leading

A young soldier named PVT Bell, new to the unit, approaches his team leader, SGT Adams, and says, "I have a problem I'd like to talk to you about."

The team leader makes time—right then if possible—to listen. Stopping, looking the Soldier in the eye, and asking, "What's up?" sends many signals: I am concerned about your problem. You're part of the team, and we help each other. What can I do to help? All these signals, by the way, reinforce Army values.

PVT Bell sees the leader is paying attention and continues, "Well, I have this checking account, see, and it's the first time I've had one. I have lots of checks left, but for some reason the PX [post exchange] is saying they're no good."

If a squad leader doesn't check, and the guy on point has no batteries for his night vision goggles, he has just degraded the effectiveness of the entire unit.

Company Commander, Operation Desert Storm

Critical Thinking

Why are the abilities to listen, understand, and communicate with empathy at least as important as technical or tactical skills? Why is it said that "listening is leading?"

SGT Adams has seen this problem before: PVT Bell thinks that checks are like cash and has no idea that there must be money in the bank to cover checks written against the account. SGT Adams, no matter how tempted, doesn't say anything that would make PVT Bell think that his difficulty was anything other than the most important problem in the world. He is careful to make sure that PVT Bell doesn't think that he's anyone other than the most important Soldier in the world. Instead, SGT Adams remembers life as a young Soldier and how many things were new and strange. What may seem like an obvious problem to an experienced person isn't so obvious to an inexperienced one. Although the Soldier's problem may seem funny, SGT Adams doesn't laugh at the subordinate. And because nonverbal cues are important, SGT Adams is careful that his tone of voice and facial expressions don't convey contempt or disregard for the subordinate.

Instead, the leader listens patiently as PVT Bell explains the problem; then SGT Adams reassures PVT Bell that it can be fixed and carefully explains the solution. What's more, SGT Adams follows up later to make sure the Soldier has straightened things out with the bank.

A few months later, a newly promoted PFC Bell realizes that this problem must have looked pretty silly to someone with SGT Adams's experience. But PFC Bell will always remember the example SGT Adams set. Future leaders are groomed every day and reflect the behavior of their past leaders. By the simple act of listening and communicating, SGT Adams won the loyalty of PFC Bell (FM 22-100).

CONCLUSION

Good discipline, excellent morale, and superior results are no accident. They also don't occur naturally. The performance of Soldiers and units in the field mirrors how rigorously they train. Responsibility for that training falls on you, a member of Army's direct, first-line leadership. Subordinates expect you to show them the Army standard and train them to achieve it. They expect you to lead by example as one of the Army's technical experts and best teachers.

Broad skills competence and strong character inform good training and good decision making. By learning as much as you can about your job, your people, and yourself, you'll Be a leader of character, Know the necessary skills, and Do the right thing. Your reward will be a team with the skill, trust, and confidence to succeed—in any situation, anywhere.

Learning Assessment

1. Describe the four key Army leadership skills.

2. Explain how competence in leadership skills grows out of the three key leadership attributes—mental, physical, and emotional.

3. What are three technical skills you think will be important for you to develop as a leader?

4. What are three tactical skills you will need to teach members of your unit?

Key Words

direct leadership
competence
tactics
fieldcraft

References

DA PAM 600-65, *Leadership Statements and Quotes*. 1 November 1985.

Field Manual 22-100, *Army Leadership: Be, Know, Do*. 31 August 1999.

ARMY LEADERSHIP— ACTIONS

Key Points

1 The Three Categories of Leader Actions

2 The Nine Specific Leader Actions Identified by Category

> Leadership is the art of getting someone else to do something you want done because he wants to do it.
>
> GEN Dwight D. Eisenhower

Introduction

The best training in the world can't make you an effective, respected leader. Successful leadership requires that you *Do* the appropriate thing at the appropriate time. The true value of both your character and competence discussed in previous sections shows in your actions. As an Army leader, most of your *Doing* will be making decisions and ensuring that your subordinates carry out your orders.

The Three Categories of Leader Actions

The three broad leader actions—influencing, operating, and improving—are not separate, but are part of a leadership action mindset that good leaders develop. As you become more experienced and confident as an Army officer, your decision making powers will develop and begin to seem natural to you.

The example you set by doing says more about you than any other facet of your job. Through study, practice, trial and error, and continuous improvement, you can become an effective, competent, confident leader. An upbeat, impartial, action-oriented leadership style will guide you through the challenges of military leadership.

Cultivation of this mentality will help you in both the easy and the difficult decisions you will face. Even when the odds are stacked against you, relying on levelheaded military professionalism wins battles, as SGT Alvin York discovered in France during a lopsided fight against a German infantry battalion in World War I.

SGT York

Alvin York performed an exploit of almost unbelievable heroism in the morning hours of 8 October 1918 in France's Argonne Forest. He was now a corporal (CPL), having won his stripes during combat in the Lorraine. That morning CPL York's battalion was moving across a valley to seize a German-held rail point when a German infantry battalion, hidden on a wooded ridge overlooking the valley, opened up with machine gun fire. The American battalion dived for cover, and the attack stalled. CPL York's platoon, already reduced to 16 men, was sent to flank the enemy machine guns.

As the platoon advanced through the woods to the rear of the German outfit, it surprised a group of about 25 German soldiers. The shocked enemy offered only token resistance, but then more hidden machine guns swept the clearing with fire. The Germans dropped safely to the ground, but nine Americans, including the platoon leader and the other two corporals, fell dead or wounded. CPL York was the only unwounded leader remaining.

CPL York found his platoon trapped and under fire within 25 yards of the enemy's machine gun pits. Nonetheless, he didn't panic. Instead, he began firing into the nearest enemy position, aware that the Germans would have to expose themselves to get an aimed shot at him. An expert marksman, CPL York was able to hit every enemy soldier who popped his head over the parapet.

After he had shot more than a dozen enemy [troops], six German soldiers charged him with fixed bayonets. As the Germans ran toward him, CPL York once again drew on the instincts of a Tennessee hunter and shot the last man first (so the ones in front wouldn't see the ones he shot fall), then the fifth, and so on. After he had shot all the assaulting Germans, CPL York again turned his attention to the machine gun pits. In between shots, he called for the Germans to give up. It may have initially seemed ludicrous for a lone Soldier in the open to call on a well-entrenched enemy to surrender, but their situation looked desperate to the German battalion commander, who had seen over 20 of his Soldiers killed by this one American. The commander advanced and offered to surrender if CPL York would stop shooting.

CPL York now faced a daunting task. His platoon, now numbering seven unwounded Soldiers, was isolated behind enemy lines with several dozen prisoners. However, when one American said their predicament was hopeless, CPL York told him to be quiet and began organizing the prisoners for a movement. CPL York moved his unit and prisoners toward American lines, encountering other German positions and forcing their surrender. By the time the platoon reached the edge of the valley they had left just a few hours before, the hill was clear of German machine guns. The fire on the Americans in the valley was substantially reduced and their advance began again.

CPL York returned to American lines, having taken a total of 132 prisoners and putting 35 machine guns out of action. He left the prisoners and headed back to his own outfit. Intelligence officers questioned the prisoners and learned from their testimony the incredible story of how a fighting battalion was destroyed by one determined Soldier armed only with a rifle and pistol. Alvin C. York was promoted to sergeant and awarded the Medal of Honor for this action. His character, physical courage, technical competence, and leadership enabled him to destroy the morale and effectiveness of an entire enemy infantry battalion.

Critical Thinking

How much of SGT York's heroism was gut instinct and how much was planned leadership action? Which leadership actions did York rely on most?

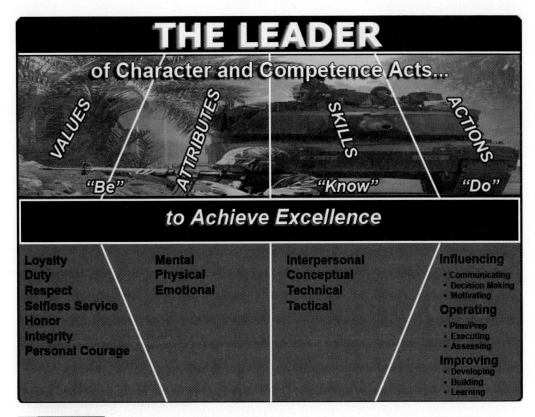

Figure 4.1 The Army Leadership Framework

The Nine Specific Leader Actions Identified by Category

Influencing Actions

An effective leader acts competently and confidently. By putting your education, training, and character into action, you set the operational attitude for your entire unit—day-to-day, task-to-task, even minute-to-minute. Remember that optimism, a positive outlook, and a sense of humor are contagious, especially when you have to make unpopular decisions.

The most important influence you have on your team is the example you set. Your actions say more about what kind of leader you are than anything else. Your Soldiers watch you all the time, so act as if you are always on duty. This *influence* you exert on your command is the first category of leader actions you will read about in this section. Each of the specific leader actions identified below supports how you *influence* people while achieving a goal or accomplish a given mission.

Communicating

To build trust, relieve stress, and allow subordinates to accomplish the mission, good leaders keep their subordinates well informed. By informing your team of your decisions—and, as much as possible, the reasons for them—you show that you value them as members of your unit. Remember that sharing accurate, timely information is the best way to manage rumor control.

As a leader, you are responsible for making sure your subordinates understand both the *substance* and the *intent* of your orders. If you have any doubt that your subordinates understand your communication, check to make sure you've made yourself clear. If there's

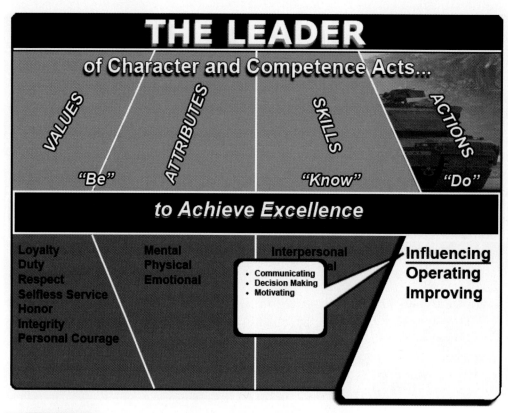

Figure 4.2 Direct Leader Actions—Influencing

back-brief

verbal or written feedback from the receiver of communication to the sender, indicating that the receiver has understood—or misunderstood—the substance and intent of the communication

troop leading procedures (TLP)

leaders at company level and below follow the troop leading procedures (TLP), which are designed to support solving tactical problems

time, even if you think your Soldiers understand, check anyway by asking for a **back-brief**. That's how you create the most effective kind of interaction: two-way communication.

If you establish a climate of trust and have trained your subordinates in how and why decisions are made, they are more likely to be willing to act on their own with a minimum of guidance.

Leaders find out what their team members are thinking, saying, and doing by using the most important communication tool: *listening*. Good leadership involves coaching, listening, teaching, and clarifying.

Decision making

Decision making is recognizing *whether* to decide, then knowing *when* and *what* to decide. It includes understanding the consequences of your decisions. Army leaders usually follow one of two decision-making guidelines: **troop leading procedures (TLP)** and the **military decision making process (MDMP)**. You will study these guidelines in detail later in your ROTC career.

Critical Thinking

How can a positive leadership mindset transform decision making into a comfortable and even pleasurable habit?

As a leader, you must set priorities. You need the personal courage to say which tasks are more important than others. In the absence of a clear priority, you must set one. Not everything can be the top priority—and you can't make good decisions without setting priorities.

As you gain leadership experience, some of the decisions you find difficult now will become easier. But there will always be difficult decisions that require imagination, rigorous thinking and analysis, and sometimes even your "gut" reaction. These are the tough decisions—in the words of one first sergeant, the ones you're getting paid to make.

With some decisions, you will need to take into account your knowledge, your intuition, and your best judgment. Remember also that any decision you make must reflect Army values. Your superiors and your Soldiers expect you as a leader to make decisions that are ethical as well as practical.

Motivating

Motivation grows out of Soldiers' confidence in themselves, their unit, and their leaders. This confidence comes from hard, realistic training and constant reinforcement. Remember that trust, like loyalty, is something your Soldiers give you only when you demonstrate that you deserve it.

You empower subordinates when you train them to do their jobs, give them the necessary resources and authority, get out of their way, and let them work. This statement of trust in your subordinates is one of the best ways to develop them as leaders. Coach and counsel them, both when they succeed and when they fail. Give positive reinforcement and credit to those who deserve it—you'll be amazed at the results.

Of course, not everyone performs to standard. In fact, some subordinates will require punishment. Using punishment to correct undesirable behavior is difficult but is a way of motivating. And sound judgment must guide you when administering punishment. Punishment is most effective when the subordinate clearly understands the infraction and sees that the punishment fits the deed. Because of legal aspects sometimes associated with it, you may need to consult with your superior officer, Army lawyer, or Army regulations before administering punishment.

Operating Actions

You are operating when you're working to get today's job done. In practice, you often must "multi-task" using different leadership values, skills, and actions. This overlapping interaction strengthens both your individual performance and that of your team. The specific leader actions identified below will allow you to accomplish missions that require your immediate attention.

Planning

A plan is a map for completing a command decision or project. Planning begins with a specified or implied mission. A *specified* mission is issued by your superior officer or from your higher headquarters. An *implied* mission is something within your area of responsibility that you decide needs to happen and for which you develop a plan of action on your own.

You may often find the *reverse planning* method useful. That's a plan that begins with the goal in mind. Start with the question: "Where do I want to end up?" and then work backward from there until you reach where you are right now.

Along the way, you should allocate resources using these familiar questions as your guide: *who, what, when, where,* and *why?* As you plan, consider the amount of time needed to coordinate and conduct each step. After you have plotted the route to the goal, put tasks in sequence, set priorities, and determine the schedule. Make sure that events are in logical order and that you have allotted enough time for completing each.

military decision making process (MDMP)

leaders at battalion level and above follow the military decision making process (MDMP), which is designed for larger organizations with staffs that help the commander make and implement decisions

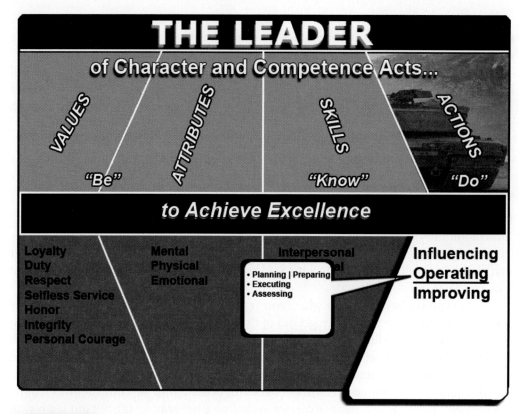

Figure 4.3 Direct Leader Actions—Operating

While leaders plan, subordinates prepare. Develop your plan while your team prepares: You'll give them advance notice of the task or mission by issuing what's called a *warning order* (*WARNO*). Based on this guidance, subordinates can then draw ammunition, for example, or rehearse key actions, inspect equipment, conduct security patrols, or begin movement while you complete the plan. Rehearsal is also an important element of preparation. Rehearsing key combat actions lets your subordinates see how things should work and builds their confidence in your planning.

If you hit an obstacle in planning, ask others with more experience for advice or clarification. Go to your superior officer only with problems you can't fix—not those for which solutions are merely difficult.

Executing

Executing means acting to accomplish the mission, moving to achieve your goals as outlined in your plan and your orders—to standard and on time. In executing your plan, at times it may seem as if everything is working against you. Adapt and improvise. Keep in mind your superiors' intent and the mission's ultimate goal. Remember that a well-trained organization accomplishes its mission, even when things go wrong.

In a tactical setting, you must know the intent of your commander two levels up. During execution, position yourself to best lead your team, initiate and control the action, get others to follow the plan, react to changes, keep your Soldiers focused, and work to accomplish the mission to Army standards.

By such actions, 1LT Christopher Dean led his Soldiers to the rescue of an ambushed patrol in Baghdad in April 2004.

1AD Lieutenant Earns Silver Star

November 23, 2004—Like many Soldiers honored as heroes, 1st Lt. Christopher Dean of V Corps' 1st Armored Division says he was just doing his job the day he earned a Silver Star for leading the rescue of a patrol ambushed in Baghdad.

"People don't say, 'I'm going to try to win a Silver Star today.' We go out and we're put in an extraordinary position, and the right people recognize what we are doing," said Dean. "I wouldn't say I was in the right place at the right time, but I guess I was fortunate to be in the wrong place at the right time."

Dean, a platoon leader in the division's Company C, 2nd Battalion 37th Armor, based in Friedberg, Germany, was helping to hand authority for the division mission over to the incoming 1st Cavalry Division at that "right time"—April 4. More important, the lieutenant's assignment that day was to serve as Quick Reaction Force tank platoon leader, with oversight for the "wrong place"—Sadr City, arguably the most violent section of Baghdad.

A patrol from 1st Cavalry was ambushed in the city. Dean rolled out immediately with four tanks under his charge. Traveling at top speed, they headed to the grid coordinates given by the besieged patrol. As soon as they arrived, the QRF was hit by a barrage of gunfire.

"We had rounds coming in from everywhere, said Dean. "It sounded like Rice Krispies popping." One of his Soldiers was killed.

Dean then led a seven-tank attack back into the engagement area to find the ambushed patrol. The .50-caliber machine gun was taken out by enemy fire, leaving him atop the vehicle with only his M4 rifle. He was hit by shrapnel from a rocket-propelled grenade blast.

Reaching the ambushed patrol, the QRF dismounted to help get the patrol out. Under heavy enemy fire they pulled out the dead and wounded and put them inside the tanks, then used one of Dean's tanks to push two damaged vehicles out of the area.

Dean's team rescued 19 Soldiers from the ambush.

In a huge ceremony Oct. 7, 1st Armored Division welcomed its "Iron Soldiers" home from their 15 months in Iraq, and Dean stood before thousands of his fellow troops as a Silver Star was pinned to his uniform, to wear along with the Purple Heart he had been presented earlier (US Army, 2004).

Assessing

Remember that Army leaders don't set the minimum standards as a goal. You must know, communicate, and enforce standards. Explain the standards that apply to your mission or project and give your subordinate leaders authority to enforce them. Then hold your subordinates responsible for achieving them. Whenever you talk about accomplishing the mission, always include the phrase "to standard." Of course, encouraging Soldiers to *exceed* the standard is a morale booster and improves results.

Army leaders constantly check things: people, performance, equipment, and resources. They check things to ensure the organization is meeting standards and moving toward the goal the leader has established. Praise good performance and figure out how to fix poor performance.

Successful assessment begins with getting an early picture of your unit's performance. The **after action review (AAR)** can help improve unit assessment and performance. Another tool Army leaders use to assess and improve is the in-process review (IPR). IPRs give leaders and subordinates a chance to talk about how the mission or project is unfolding. The IPR enables you to catch problems early and correct or avoid them. Once the mission is underway, think of an IPR as a refueling stop on the way to the mission objective.

Good leaders provide immediate, honest, direct assessment to their subordinates. Tell them about their strengths and let them know where they can improve. Ask your team members to create their own individual plans of action for self-improvement. They should see your assessment of their performance as a positive experience and a chance for improvement—perhaps even a step toward promotion. They should value your feedback as a beneficial opportunity for them to tap into your experience and expertise.

Follow up your missions and training with AARs so that subordinates can talk about what they did well, what they could have done better, and what they should do differently next time.

Improving Actions

Simply put, improving actions are what you do to leave your unit better than you find it. Improving actions fall into the categories highlighted in Figure 4.4: developing, building, and learning. Each of these skills will allow you to improve the Army—both its organizations and people—and meet long-term goals.

Developing

Developing refers to improving people. You improve your unit and the entire Army when you develop your subordinates. It's your duty as an officer to invest your time and energy to help them reach their full potential. This principle applies not only to all ranks and levels of Army Soldiers but also to Department of the Army civilians and the Army reserve forces.

The Army is becoming increasingly technologically complex. The Contemporary Operating Environment demands ever-more sophisticated and better-trained Soldiers. Developing yourself and your subordinates involves several areas: self-development, teaching, developmental counseling, mentoring, and coaching.

Self-development is a process to increase your own readiness and potential for positions of greater responsibility. You can use the dimensions of the Army leadership framework to help you determine which areas to develop. Self-development is continuous and takes place during institutional training and operational assignments. Self-development is a joint effort by you, your first-line leader, and your commander.

Teaching is passing on knowledge and skills to subordinates. To be an Army leader, you must be a teacher—a primary task for first-line leaders. Teaching focuses primarily on technical and tactical skills, and technical competence is essential to effective teaching. To

after action review (AAR)

a briefing following a mission or training that covers best practices and lessons learned— subordinates are able to give feedback to superiors on the positive and negative aspects of the team's performance

After-Action Reviews (AARs)

Army leaders use AARs as opportunities to develop their subordinates. During an AAR, the leader gives the team a chance to talk about how they think the training or mission went. AARs center on constructive, useful feedback about what the team did well and what it can do better next time. AARs give leaders a chance to hear what's on their subordinates' minds. (FM 7-1 and TC 25-20 discuss how to prepare, conduct, and follow up after AARs.)

Critical Thinking

What is the value of habitually using the in-process review (IPR) and after action review (AAR)? How do these assessment techniques affect your team members' understanding of your expectations?

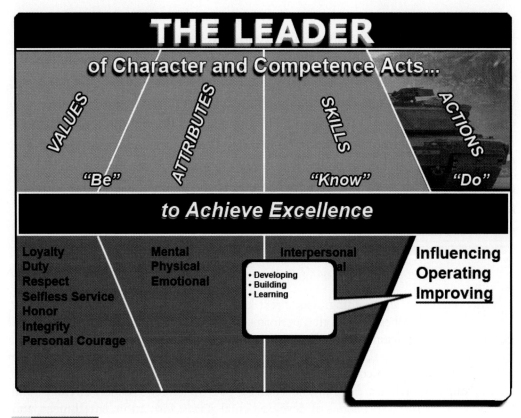

Figure 4.4 Direct Leader Actions—Improving

develop subordinates, you must be able to demonstrate the technical and tactical skills you expect them to demonstrate.

Developmental counseling is subordinate-centered communication to plan the actions necessary for subordinates to achieve their individual or unit goals. It should focus on today's performance and problems as well as tomorrow's plans and solutions. In effective developmental counseling your subordinate takes an active role—listening, asking for feedback, seeking out what the counselor has to say. Be honest and professional with your subordinates during developmental counseling.

Mentoring is the development of each subordinate through observing, assessing, coaching, teaching, developmental counseling, and evaluating the results in your Soldiers. Mentoring is an inclusive process (not an exclusive one) for everyone under your command. Mentoring begins with your setting the right example and shows your subordinates a model of values, attributes, and skills in action.

Coaching involves a leader's assessing performance based on observations, then helping the subordinate develop an effective plan of action to sustain strengths and overcome weaknesses. Less formal than teaching, coaching opportunities occur all the time—even in informal or social situations—when you concentrate on developing potential leaders.

In his 1995 autobiography, GEN Colin Powell wrote of the benefits of coaching he received while serving as a lieutenant in Germany in 1959-60:

The Lieutenant's Temper

Captain William C. Louisell [the new company commander] was a West Pointer and a former tactics instructor at the military academy....

One day I was in the orderly room on the phone, shouting at a fellow lieutenant at the top of my lungs, when Louisell walked in. He took me aside and chewed me out for my behavior. Shortly afterward, I received my efficiency report. To the layman, it might not seem disastrous. Louisell had said of me, "He has a quick temper which he makes a mature effort to control." But in the code of efficiency report writing, I had taken a hit. These words marked the only negative comment on my performance since the first day I had put on a uniform in ROTC. Louisell called me in, sat me down, and raised the matter of the blowup on the phone. "Don't ever show your temper like that to me or anyone else," he warned. It was demeaning to everybody. I still have a hot temper. I still explode occasionally. And whenever I do, I hear Bill Louisell's warning voice (Powell, 1995).

Building

Building refers to team building. As a direct leader, you improve your organization by building a strong, cohesive team that performs to standard, even in your absence. If the leaders of the small teams that make up the Army are competent, and if their members trust one another, those teams and the larger organization will hang together and get the job done. Additionally, a cohesive team accomplishes the mission much more efficiently than a group of individuals can. But teams don't appear by accident. Leaders carefully build and guide them through a series of developmental stages: formation, enrichment, and sustainment.

Formation stage. Teams work best when new members are oriented quickly, so that they're made to feel a part of the team. Two steps—reception and orientation—are dramatically different in peace and war. In combat, this sponsorship process can literally mean life or death to new members and to the team. Reception is the leader's welcome. Orientation begins with meeting other team members, learning the layout of the workplace, learning the schedule and other requirements, and generally getting to know the operating environment.

In combat, unit leaders may not have time to spend with new members. In this case, new arrivals are often paired up with a "buddy" who will help orient them until they've learned their way around.

Enrichment stage. New teams and new team members gradually move from questioning everything to trusting themselves, their peers, and their leaders. Leaders earn that trust by listening, following up on what they hear, establishing clear lines of authority, and setting standards. One of most important things you do as a leader is strengthening your team's training. Training molds a group of individuals into a team while preparing them to accomplish their mission. Training occurs during all three team building stages, but is particularly important during enrichment as this is the point the team is building its collective skills.

Sustainment stage. At this stage, members think of the team as "their team." They own it, have pride in it, and want the team to succeed. At this stage, team members will do what needs to be done without being told. Every new mission gives the leader a chance to make the bonds even stronger, to challenge the team to reach for new successes. He or she continues to train the team to maintain proficiency to accomplish its missions. Finally, the leader works to keep the team going in spite of the stresses of and losses in combat.

As an Army leader, you are the ethical standard-bearer for your unit. You're responsible for building an ethical climate that demands and rewards behavior consistent with Army values. Leaders who create a healthy ethical environment inspire confidence in their subordinates. That, in turn, builds the unit's will to succeed.

Learning

Learning refers to your Soldiers' self-development. Establish a team attitude that rewards collective learning and ensures that your subordinates learn from their collective experiences.

Whenever you have people doing difficult jobs in a complex, pressure situation, problems arise. Mistakes are a natural part of the training process. Effective leaders use mistakes as opportunities to teach others to do things better and share what they have learned with other leaders in the organization, both peers and superiors. You need to make sure your subordinates understand which sorts of mistakes are acceptable and which are not. For example, helicopter mechanics work and train in a "zero-defect" environment; they can't allow aircraft to be mechanically unstable during flight.

GEN Powell wrote about team-building lessons from his time as a lieutenant in Germany:

Understanding GIs

My first platoon sergeant was Robert D. Edwards, from deepest Alabama....

The troops feared Edwards, and with reason. Once, I had to explain to him why he could not keep a Soldier who had gone AWOL chained to the barracks radiator. Edwards found my reasons puzzling and went off muttering about the decline of discipline. While he was feared, he was, at the same time, respected and revered by the men. They understood Edwards. He was in their corner. No matter how primitive his methods, he had one concern—the welfare of the platoon and the men in it. If they soldiered right, he looked out for them.

I came to understand GIs during my tour at Gelnhausen [in Germany]. I learned what made them tick, lessons that stuck for thirty-five years. American Soldiers love to win. They want to be part of a successful team. They respect a leader who holds them to a high standard and pushes them to the limit, as long as they see a worthwhile objective. American Soldiers will gripe constantly about being driven to higher performance. They will swear they would rather serve somewhere easier. But at the end of the day, they always ask: "How'd we do?"

And I learned what it meant when Soldiers brought you problems.... The day Soldiers stop bringing you their problems is the day you have stopped leading them. They have either lost confidence in you or concluded you do not care. Either case is a failure of leadership (Powell, 1995).

You must talk to your Soldiers.... I don't just mean in formation or groups, but one-on-one. Take time ... to really talk to a Soldier, one Soldier a day.

Command Sergeant Major Daniel E. Wright

CONCLUSION

"Reverse plan" your career as an Army leader for just a moment. In the future, what kind of leader do you aspire to be? Are your subordinates able to understand a situation the way you do? Do they anticipate outcomes because you've trained them to look at situations with reason, ethics, and an action-oriented attitude?

Some actions are easy to take; some require problem solving and more involved decision making. As you gain experience as a leader, some of the decisions you find difficult now will become easier, and you'll become more comfortable in your command role.

Any decision you make must reflect the Army values discussed earlier. The Army has established standards for all military actions. You must know, communicate, and enforce those standards.

Your responsibility as a leader requires that you manage subordinates face-to-face as they carry out their duties. By living Army values and providing the proper role model for subordinates, you will increase your skill in building strong, cohesive teams and establishing effective command.

Learning Assessment

1. What are the three categories of Army leadership actions?
2. Explain how the nine leadership actions form a decision making mindset.
3. What are the essential differences between Army values, skills, and actions?
4. Why is your competence in carrying out Army actions important to the overall success of your team?

Key Words

back-brief
troop leading procedures (TLP)
military decision making process (MDMP)
after action review (AAR)

References

Field Manual 22-100, *Army Leadership: Be, Know, Do.* 31 August 1999.

Field Manual 7-1, *Battle Focused Training.* 15 September 2003.

Powell, C. (with Persico, J.). (1995). *My American Journey.* New York: Random House, Inc.

TC 25-20, *A Leader's Guide to After-Action Reviews.* 30 September 2003.

US Army. (23 November 2004). 1st AD Lieutenant Earns Silver Star. *Soldier Stories.* Retrieved 24 August 2005 from http://www4.army.mil/ocpa/soldierstories/story.php?story_id_key=6591

Section

1

TIME MANAGEMENT

Key Points

1 The Process for Effective Time Management

2 Barriers to Time Management

3 SMART Goals

4 The POWER Model

In the Army, we speak of resources in terms of the three M's—men, money, and materials. To these three, we must add *time* . However, there is a distinct difference between time and the three other resources. If we do not use our money or materials today, they are available tomorrow. To a lesser degree, this is also true of men. It is not at all true of time, for time is a highly perishable commodity. An hour lost today is lost forever!

GEN Bruce C. Clarke

Introduction

To succeed as a leader, you must competently manage your resources—people, materials, and information. But perhaps the most precious resource you must manage is time. You can't replace time once it's gone.

Time management can also be called "personal management" since *you* are the only thing you can control. The time-management skills you develop as a cadet and student will set a pattern for you as a future Army officer. That's why you must cultivate a process for effectively managing time *today*.

As this section points out, you will encounter significant barriers to effective time management. But you can overcome those barriers by setting goals and using systematic techniques.

The Process for Effective Time Management

Time management is a system for getting things done as efficiently and productively as possible. As you assume leadership responsibilities in the Army, your time will become even more valuable and its management more complex. You will need to manage not just your own time, but also the time of those you lead. When you put an efficient time-management system into action, you become a more effective leader, because your team members perceive the value you place on both your time and theirs. The respect you show for their time will support everyone's efforts to become more efficient as the unit works towards meeting task deadlines and completing the mission schedule.

But it all starts here in ROTC, where your college studies require significant time-management skills. Your success at cultivating good time-management skills depends on your ability to set aside enough time for each of your classes, sports, and other activities. Doing this efficiently requires that you:

1. Recognize time wasters (black holes)
2. Set goals that will reduce patterns of wasted time
3. Adopt a system like the POWER model.

The **POWER** model, discussed below, will move you towards your personal, professional, and career goals.

Barriers to Time Management

Have you ever heard of a black hole in astronomy? It's a point in space where a massive object pulls in all light near it. Nothing can escape.

Critical Thinking

Why do you think good time-management skills in an officer foster a productive work ethic among Soldiers in the officer's small Army unit? How does the value you place on time management translate into your team members wasting less time? Is there a trickle-down effect?

Your Calendar

Create your own semester calendar that includes all classes, key assignments, fitness workouts, sports events, extracurricular activities, social events, and any other important times for the coming semester. Use this calendar to determine how much time you still have open and to identify weeks (usually during midterms and finals) when you will need to plan and work ahead to avoid becoming overwhelmed.

Critical Thinking

Pick a goal you have set for yourself: Make enough money to buy a car; graduate with honors; land a summer internship at the state capital; improve your foreign language skills. Now turn it into a SMART goal using the model below. What did you learn about yourself and your goal?

In your schedule, "black holes" are spaces of time that eat into your productivity and prevent you from reaching your goals and the goals of others you work with. Black holes devour productive time and consume your efficiency.

The ability to identify black holes is the first and most important aspect of good time management. To do this, you should conduct an informal inventory of how you use your time. Consider a typical week, and, within that week, a typical day. Most people are surprised at how much of their time is unfocused and unstructured—without a specific goal or purpose. Two causes of black holes are *procrastination*—putting things off—and *distracters*—things that take us away from our planned work or activities.

One way to fight procrastination is to realize that it will only make things worse. As George H. Lorimer, editor of the old *Saturday Evening Post,* once put it, "Putting off an easy thing makes it hard, and putting off a hard one makes it impossible." Distracters are often subtle. They might be talking on the phone with friends or relatives, losing track of time while Instant Messaging or playing video games, or helping other people on their projects. Or you may face an unexpected change in your work schedule. Learning to deal with distracters—saying no when necessary; turning off your cell phone, the IM, or the video game; negotiating your schedule with your supervisor; even finding a different spot to work or study away from distractions—builds the discipline that helps you stay on course.

Trying to overcome these two sources of black holes in your life goes a long way toward improving your use of available time. Additionally, the better you get at completing work on schedule, the better you'll get at scheduling time for recreation, hobbies, social activities, and other things you enjoy.

SMART Goals

Goal setting is a critical part of managing your time. If you don't know where you are going, how can you possibly calculate how long it will take you to get there?

TABLE 1.1	The **SMART** Model
Specific	make the goal concrete and clear
Measurable	decide how you will measure success in reaching the goal
Achievable	keep goals reasonable—milestones are helpful
Realistic	consider other factors that may affect the goal
Time-bound	make yourself accountable for a specific date

TABLE 1.2	The **POWER** Model
Prepare	set SMART goals
Organize	keep a calendar to know where you are, where you've been, and what's ahead
Work	work on establishing boundaries for your time—prioritize activities, avoid procrastinating, learn when to say no, and keep track of how long important tasks take
Evaluate	review how you're spending your time
Rethink	explore better and better ways of managing your time—identify what you need to change

The **SMART** model for goal setting, spelled out in Table 1.1, is a useful starting point for filling up black holes with useful activities—activities that get you where you want to go on your mission, in your career, and in your life.

The POWER Model

Once you have established your goals, you need to apply an effective time-management system to reach them. The **POWER** model outlined in the table above provides you with such a system. Practice techniques of good time management, such as completing tasks before moving to the next item or limiting distractions when doing exercise or homework. Some useful tools include a calendar, a weekly schedule, and a To-Do list. You can find sophisticated time-management aids at an office-supply store or perhaps your student bookstore. Your personal computer may also have useful time-management software.

In applying any time-management strategy, it's a good idea to keep in mind that you must be flexible. Since no one can foresee the future, you need to be willing to modify your plans to accommodate events and even a few surprises.

Tools for Productive Time Management

1. Calendar

2. Weekly Schedule

3. To-Do List

Time management is a system for getting things done as efficiently and productively as possible.

CONCLUSION

Successful leaders such as GEN Bruce Clarke recognize that using time more efficiently than the enemy is the first requirement for victory. They also realize that successful time management is a never-ending process.

As a cadet and future Army leader, you must adopt time-management systems that allow you to envision how to achieve goals and objectives, plan to reach those goals, organize your time, and continue to improve.

It's time to take control of your time using the SMART technique and POWER model you've learned. Start today!

Learning Assessment

1. Trade one of your personal goals for this semester with a fellow cadet in the class. How specific is his or her goal? Is it measurable? Achievable? Realistic? What is the timetable for achieving the goal? Brief your partner on your SMART observations about his or her goal.

2. Describe the POWER model for effective time management. List some of the barriers you might encounter and how you would plan to overcome them.

Key Words

SMART
POWER

References

Clark, B. C. (1973). *Guidelines for the Leader and Commander*. Harrisburg, PA: Stackpole Books.

Ellis, D. (2003). *Becoming a Master Student* (10th Ed.). Boston: Houghton Mifflin Company.

Hughes, R. L., Ginnett, R. C., & Curphy, G. J. (1999). *Leadership: Enhancing the Lessons of Experience* (3rd Ed.). New York: Irwin/McGraw-Hill.

Section 2

HEALTH AND FITNESS

Key Points

1 Components of Fitness

2 Principles of Exercise

3 Frequency, Intensity, Time, Type

4 Safety and Smart Training

5 Nutrition and Diet

To every man there comes in his lifetime that special moment when he is figuratively tapped on the shoulder and asked to do a very special thing—unique to him and his talents. What a tragedy if that moment finds him unprepared or unqualified for that work.

Sir Winston Churchill

Introduction

Have you ever noticed during sports competition that the individual or team that tires first often loses? It's the same for Soldiers. Your ability to cope with battlefield challenges depends greatly on your level of physical fitness. Physical fitness not only determines how well you perform in combat, but also enhances your overall quality of life, improves your productivity, and brings about positive physical and mental changes.

Your physical fitness benefits both the Army and you. The Army needs physically fit Soldiers, and when you are fit, you are more likely to lead an enjoyable, productive life.

As an officer, how important is your level of physical fitness? How does your fitness affect your unit's combat readiness?

You're probably tired of hearing how important it is to be in great shape as an officer, but it's a basic truth. You don't have to be the best at everything, but you definitely need to be one of the most physically fit Soldiers in your platoon. Morale improves when your Soldiers are constantly trying to beat you in a run or in an individual event like pull-ups or the rope climb. When that happened to me as a platoon leader that meant instant respect. My Soldiers all knew I could run or road march to the end with any of them. You can't motivate Soldiers in a road march if you are visibly in bad shape. It's a sad [sight] when a lieutenant can't lead his Soldiers physically. Ask yourself: How can you lead or motivate your Soldiers if you're not at the head of the formation? When it comes to the combat environment, physical fitness is crucial. If you allow your Soldiers to deploy in poor condition, you have failed them. Being physically fit out here [in Afghanistan] will help your Soldiers bear some of the rigors of a combat tour: less sleep, very random and increasingly changing rest patterns, extreme heat, heavy weights, and less than standard nutrition, to name a few. Staying in shape in a combat environment can be a tough task, because you may lack the time or facilities to exercise as you may have been able to in garrison. However, solid cardiovascular fitness will make a significant difference in higher elevations, and upper body and leg strength may prevent exhaustion from heavy gear in hot weather (3rd Brigade, 25th Infantry Division (L), 2005).

ILT Eliel Pimentel

Components of Fitness

Your physical fitness is your ability to perform physical work, training, and other activities throughout your daily work schedule. Physical fitness is multidimensional, and—based on your goals—some components will be more valuable than others.
Five key components define your physical fitness:

- *Cardio-respiratory (CR) endurance*—how efficiently your body delivers oxygen and nutrients for muscular activity and transports waste from the cells.
- *Muscular strength*—the greatest amount of force your muscle or muscle group can exert in a single effort.
- *Muscular endurance*—the ability of your muscle or muscle group to perform repeated movements for extended periods.
- *Flexibility*—the ability to move your joints (elbow or knee, for example) or any group of joints through their entire normal range of motion.
- *Body composition*—the amount of body fat you have in comparison to your total body mass.

Improving the first three of these components will improve your body composition by decreasing your body fat. Excessive body fat detracts from the other fitness measures, reduces your physical and mental performance, detracts from your appearance, and increases overall health risks. One measurement of body fat is as a percentage of your total weight. The Army's maximum allowable percentages of body fat, by age and gender, are listed in Figure 2.1.

Besides your physical fitness, you should also work to improve your *motor fitness*. Motor fitness—speed, agility, muscle power, eye-hand coordination, and eye-foot coordination—directly affect a Soldier's performance on the battlefield. Appropriate training will improve these elements up to each Soldier's individual potential.

The goal of the Army's fitness program is to improve physical and motor fitness through sound, progressive, mission-specific physical training at both the individual and unit levels.

Body Fat Standards				
Ages	17–20	21–27	28–39	40+
Males	20%	22%	24%	26%
Females	28%	30%	32%	34%

Figure 2.1 Body Fat Standards

Principles of Exercise

P-R-O-V-R-B-S

the basic exercise principles—Progression, Regularity, Overload, Variety, Recovery, Balance, and Specificity

Practicing the basic exercise principles is crucial for you to develop an effective fitness-training program. The principles of exercise apply to everyone at all levels of physical training, from the Olympic champion to the weekend golfer. They apply especially to fitness training for military personnel, because having standard fitness principles across the organization saves time, energy, resources—and prevents injury.

You can easily remember the basic principles of exercise if you recall the P-R-O-V-R-B-S acronym:

P *Progression*—The intensity and duration of exercise must gradually increase to improve your fitness level. A good guideline for improvement is a 10 percent gain at specified intervals.

R *Regularity*—To achieve effective training you should schedule workouts in each of the first four fitness components at least three times a week. Regularity is also key in resting, sleeping, and following a good diet.

O *Overload*—The workload of each exercise session must exceed the normal demands placed on your body to bring about a training effect. You've often heard this expressed as "No pain, no gain." A fitness trainer, such as your ROTC instructor, can help you learn to tell the difference between pain that results from an optimum level of overload and pain that indicates potential injury.

V *Variety*—Changing activities reduces the boredom and increases your motivation to progress.

R *Recovery*—You should follow a hard day of training for a given component of fitness by an easier training or rest day for that component. This helps your body recover. Another way to promote recovery is to alternate the muscle groups you exercise every other day, especially when training for strength and muscle endurance.

B *Balance*—To be effective, a fitness program should address all the fitness components, since overemphasizing any one of them may detract from the others.

S *Specificity*—You must gear training toward specific goals. For example, Soldiers become better runners if their training emphasizes running drills and techniques. Although swimming is great exercise, it will not improve a two-mile-run time as much as a coordinated running program does.

Frequency, Intensity, Time, Type (FITT)

To succeed in any fitness-training program you undertake, you must track your frequency, intensity, time, and type of exercise (FITT). You can use the acronym **FITT** to remember these factors easily. While FITT is just one method of developing a proper long-term physical fitness regime, fitness experts agree that you need these factors to have an effective, safe daily workout program.

Frequency

Frequency is the number of workouts you perform each week. A basic guideline is three to five cardiovascular workouts, two to three strength workouts, two to five calisthenics workouts and three to six flexibility workouts weekly.

Intensity

Intensity is how hard you work out. You can measure intensity by something called RPE (Rating of Perceived Exertion), which is a psychological scale and reflects how hard the workout feels to you.

The most commonly used indicator of your workout intensity is your heart rate. Ideally, you should stay within a productive heart-rate zone. You can use your age to find your Target Heart Rate (THR).

Finding Your THR

- Your maximum heart rate (MHR) is approximately 220 minus your age
- Your lowest target heart rate is equal to MHR × .60
- Your highest target heart rate is equal to MHR × .85.

FITT

the factors of a successful fitness-training program: frequency, intensity, time, and type

Rating of Perceived Exertion–Two RPE scales are in common use. The scales are either 6 to 20 or 0 to 10. Although the RPE scale of 6–20 does not measure heart rate, it theoretically correlates (for example: 6=60 heartbeats per minute, or bpm; 7 = 70bpm; 20 = 200bpm). Your RPE on the 6-20 scale should be between 12 and 16.

Your THR zone is between the lowest and highest THR calculated above. As you begin your exercise routine, your heart rate should be on the lower end of your THR zone. Exercising above the zone increases your risk of injury and reduces your ability to perform optimally.

Easy Versus Hard

Exercise in moderation. Never exercise a particular muscle group hard (at a high intensity or for a long time) two days in a row. You should always follow a hard workout with a light day or a day off. For the best development, more is not always better.

Time

Like intensity, the *time* you spend exercising depends on the type of exercise you are doing. At least 20 to 30 continuous minutes of intense exercise will improve cardio-respiratory endurance. For muscular endurance and strength, exercise time equals the number of repetitions you do. For the average person, eight to 12 repetitions with enough resistance to cause muscle failure will improve both muscular endurance and strength. As you progress, you will make better strength gains by doing two or three sets of each resistance exercise.

Use flexibility exercises or stretches for varying times, depending on the objective of the session. While warming-up before a run, for example, hold each stretch for 10 to 15 seconds. To improve flexibility, stretch during your cool-down as well, holding each stretch for 30 to 60 seconds. If flexibility improvement is your goal, devote at least one session per week to developing that component.

Type

Type refers to the kind of exercise you perform. When choosing the type, consider the principle of specificity. Some people overemphasize cross training and you should avoid this pitfall. For example, to improve your level of CR fitness (the major fitness component in the two-mile run), do CR types of exercises. The basic rule is that to improve performance, you must practice the particular exercise, activity, or skill you want to improve. For example, to be good at push-ups, *you must do push-ups*. No other exercise will improve push-up performance as effectively.

Safety and Smart Training

Before you begin an exercise program, ask your physician to give you a checkup. Your doctor can advise you to avoid or participate in activities based on your current health and history. Be sure to stay within your limits. If you are injured while exercising, remember to **P-R-I-C-E** your recovery.

P *Protect*—Protect the injured area from further injury. You can wrap it lightly in an elastic bandage or wear a padded brace. Do not tightly or heavily tape up an injury, as good circulation is important to healing.

Critical Thinking

Describe a stressful physical event you have experienced (in training, sports, work, or school). How might improved physical fitness have helped you?

R *Rest*—Rest the injured area. Use a sling, cane, brace, or crutch as necessary to take your weight and decrease activity off the affected body part. Keep the joint or muscle as inactive as possible.

I *Ice*—Apply ice to the injured area for five to 15 minutes. Wrap several handfuls of crushed ice in a towel and hold it on and around the injured area. Many people instinctively try to soak an injury in warm water, and while this increases blood flow to the injury, it does not ease the inflammation and swelling.

C *Compression*—Wrap an elastic bandage around the ice to compress the injured area lightly—but not enough to cut off circulation to the injured area. After the cold compress, wrap the affected area lightly in an elastic bandage or use a flexible brace. Don't wrap any injury too tightly, as this will cut off good circulation to the injury.

E *Elevation*—Raise the affected area slightly to reduce swelling and inflammation.

In addition to P-R-I-C-E, you can talk to your doctor about using anti-inflammatory medication as needed, such as aspirin, acetaminophen, or ibuprofen. You should check to see if you have allergies to these drugs before use. *Under no circumstances* should you take them while drinking alcohol.

Smart Training

You live in your skin and know how your body feels and works best. That is why you should take responsibility for managing your own fitness-training program. Knowing your limits and capabilities is key to setting goals for physical fitness improvement.
Smart training means observing some well-recognized guidelines:

- *Progression*—As you have seen, increasing intensity and/or duration by 10 percent at regular intervals is a good idea.
- *Warm-up*—Always take a few minutes to warm up your muscles to reduce your chances of injury. Your warm up should include some running in place or slow jogging, stretching, and calisthenics. It should last five to seven minutes and should occur just before the CR or muscular endurance and strength part of the workout.
- *Stretching*—Critical to improving your flexibility, stretching increases your overall fitness and reduces the chance of muscle injury. After exercising, you should cool down by walking and stretching until your heartbeat reaches 100 bpm and heavy sweating stops.
- *Mechanics*—Concentrate on your form when exercising. Maintain intensity levels, but don't let your form suffer. You will not improve by doing exercises or repetitions incorrectly—you only increase your chances of injury.
- *Healthy Diet*—You've heard that "You are what you eat." Food is your source of strength and energy. What you eat will dramatically affect your ability to maintain and improve your overall fitness.

Nutrition and Diet

Complete physical fitness is not just about exercise, but also includes good nutrition and a sensible diet. You maintain a healthy body weight and body fat percentage through sound diet and exercise to ensure the best health, fitness, and physical performance. All of these things are relevant to maintaining military readiness and achieving peak performance.

The Food Pyramid

Knowing the US Department of Agriculture (USDA) Guidelines and understanding the **Food Pyramid** to determine your daily requirements of carbohydrates, proteins, and fat will help you make healthy food choices and improve your physical fitness. A new version

Food Pyramid

an Agriculture Department nutrition tool to help you choose the foods and amounts right for you

of the pyramid debuted in 2005 and shows the types of foods and the proportions that most healthy people should eat.

In addition, USDA has an interactive website to help you track your diet. Visit *www.MyPyramid.gov*, where you can personalize your diet by age, gender, and general fitness level.

A healthy diet has the right kinds of foods in the right amounts. Look at the Food Pyramid in Figure 2.2. The person walking up the steps on the left represents the need for daily physical activity and different individuals' different nutrition needs. The different widths of the food group bands indicate the need for proportion—how much you should choose from each group.

The six color bands symbolize the food you need daily from each group for good health.

- Orange (grains): USDA recommends you eat at least three ounces of whole-grain bread, cereal, crackers, rice, or pasta every day. Half your grains should be whole. To make sure you're eating whole grains, look for the word "whole" before the grain name on the list of ingredients.
- Green (vegetables): You should vary vegetable servings, eating more dark green vegetables, orange vegetables, and dried beans and peas.
- Red (fruits): Eat a variety of fresh, frozen, canned, or dried fruit, but go easy on fruit juices, which may contain empty calories in the form of added sugars and sweeteners.
- Yellow (oils and fats): Most of your fats should come from fish, nuts, and vegetable oils. Limit your consumption of solid fats like butter, stick margarine, shortening, and lard. Instead, when possible, consume foods with omega fish oils, which help maintain your cardio-vascular health.
- Blue (milk, an important source of calcium): Choose low-fat or fat-free milk. If you don't or can't drink milk, choose lactose-free products or other sources of calcium, such as hard cheese (cheddar, mozzarella, Swiss, or parmesan), cottage cheese, and low-fat or fat-free yogurt (including frozen yogurt).

- Purple (meat, beans, and eggs): You may notice that this band, like the yellow band for oils, is thinner than the others. This visually reminds you to "Go lean on protein." Choose low-fat or lean meats and poultry that are baked, broiled, or grilled

Figure 2.2 The Food Pyramid

rather than fried. Vary your choices, including more fish, beans, peas, nuts, and seeds. If meat typically covers most of your plate, take another look at the Food Pyramid.

Substances to Avoid

Proper health and fitness reflect a mature decision you make to set a good example for your unit. Moreover, it's a wise lifestyle choice that will help you live a longer, more productive life. As an officer in training, you should avoid substances that detract from your physical performance and even harm your health (drugs, tobacco, alcohol, etc.).

Alcohol

Many people in our society have traditionally believed that alcohol—wine, beer, or hard liquor—relaxes you, increases your self-confidence, and alters your perception of stress or fatigue. It's true that for most people, light consumption of alcoholic beverages can be a pleasant social diversion. But habitual, heavy drinking or binge drinking can cause severe dehydration, decreased performance, dependence, and harm to your metabolism.

The Army expects you to exercise your judgment and drink responsibly, to include obeying all laws regarding driving and the legal drinking age, if you choose to drink at all. And never drink to "quench your thirst" before, during, or after a workout.

Tobacco

Cigarettes, cigars, and "smokeless" tobaccos contain a whole gamut of cancer-causing chemicals that provide no positive health effects. Some maintain that the "buzz" from tobacco leads to improved performance and reaction times, but no medical evidence supports this position. In the interest of good physical fitness, it is better if you don't smoke at all. If you do smoke, however, limit your intake and avoid smoking before, during, and after workouts. Smoking increases your heart rate and blood pressure.

Controlled Substances

Controlled substances are those strictly regulated by the government and may require medical prescription. You should use such substances only under medical supervision. Other drugs such as amphetamines, narcotics, steroids, and other so-called "performance enhancing drugs" are illegal and banned by the military. These drugs change performance by increasing central nervous system arousal. They increase your heart rate and blood pressure and they may cause dizziness, nausea, irritability, insomnia—even death. No one interested in good physical fitness consumes these substances; they can only detract from your performance in both the short and long term. The Army forbids their use.

You can find the Army's health-promotion and wellness website at www.hooah4health.com.

CONCLUSION

Health and fitness are integral parts of military life. They are critical for readiness and important to the well-being of the individual Soldier. Although not a cure-all, a properly planned fitness program yields many physical and mental benefits. Effective physical training can improve your body composition (decrease body fat and increase lean body weight), ability to work, mental alertness, self-confidence, and general well-being. Exercise also decreases metabolic and mental health risks such as high blood pressure, coronary heart disease, stroke, anxiety, depression, and much more.

With assistance from your ROTC instructors, you now should be able to apply the Army's general physical fitness principles to create a self-directed physical training program that meets your needs and fulfills your personal and professional goals.

Learning Assessment

1. What are the components of physical fitness?
2. Describe the principles of physical fitness as expressed by the acronym P-R-O-V-R-B-S.
3. Explain the key factors of physical fitness training (FITT).
4. Explain how you can apply the USDA Food Pyramid to make improvements in your diet.

Key Words

P-R-O-V-R-B-S
FITT
P-R-I-C-E
Food Pyramid

References

3rd Brigade, 25 Infantry Division (L). May 2005. *Operation Enduring Freedom: Afghan Leader Book, April 2004–May 2005*. Retrieved 9 August 2005 from http://rotc.blackboard.com/courses/1/CCR/content/_488714_1/Operation_Enduring_Freedon_Leader_Book_Apr_04___May_05.pdf

AR 40-501, *Standards of Medical Fitness*. 1 March 2005.

AR 350-15, *Army Physical Fitness Program*. November 1989.

AR 600-9, *The Army Weight Control Program*. 10 June 1987.

DA PAM 350-18, *US Military Academy Cadet Army Orientation Training*. 28 May 1974.

Field Manual 7-1, *Battle Focused Training*. 15 September 2003.

Field Manual 21-18, *Foot Marches*. 1 June 1990.

Field Manual 21-20, *Physical Fitness Training*. Change 1. 1 October 1998.

Section

3

INTRODUCTION TO STRESS MANAGEMENT

Key Points

1 Defining Stress

2 Causes of Stress

3 Symptoms of Distress

4 Managing Stress

5 Depression

6 Suicide

Remember that the mind and body are one and that psychological health is just as important as physical health to your overall well-being (Pawelek & Jeanise, 2004).

Health Tips from Army Medicine

Introduction

Stress is a fact of life, wherever you are and whatever you are doing. You cannot avoid stress, but you can learn to manage it so it doesn't manage you.

Changes in our lives—such as going to college, getting married, changing jobs, or illness—are frequent sources of stress. Keep in mind that changes that cause stress can also benefit you. Moving away from home to attend college, for example, creates personal-development opportunities—new challenges, friends, and living arrangements. That is why it's important to know yourself and carefully consider the causes of stress. Learning to do this takes time, and although you cannot avoid stress, the good news is that you can minimize the harmful effects of stress such as depression or hypertension. The key is to develop an awareness of how you interpret, and react to, circumstances. This awareness will help you develop coping techniques for managing stress. For example, as an Army platoon leader, managing stress will require techniques that include an awareness of yourself and your Soldiers.

As you will see, the stress you encounter as a student differs in intensity from what you may experience in the Army, particularly while deployed or in combat. The principles and techniques you use to manage stress are similar, however, as reported by this second lieutenant after returning from the war in Afghanistan:

How do you combat fatigue, stress, and fear in yourself? In your Soldiers?

In the past seven months (in Afghanistan) I have learned a lot about how I deal with combat fatigue and stress. I have found that finding a little time for myself each day or even each week allows me to regenerate and focus. Having a sense of humor and not taking things so personally have also helped reduce my stress levels. Keeping a notebook with me at all times and writing tasks, missions, or even just things to do has helped me keep my mind at ease, rather than thinking that I have forgotten to do something. Maintaining communication with my family and friends, whether through e-mail or phone conversations, also keeps me grounded . . . (3rd Brigade, 25th Infantry Division (L)).

2LT Gisela Mendonca

Defining Stress

Stress is the way human beings react both physically and mentally to changes, events, and situations in their lives. People experience stress in different ways and for different reasons. The reaction is based on your perception of an event or situation. If you view a situation negative, you will likely feel *distressed*—overwhelmed, oppressed, or out of control. Distress is the more familiar form of stress. The other form, *eustress*, results from a "positive" view of an event or situation, which is why it is also called "good stress."

Eustress helps you rise to a challenge and can be an antidote to boredom, because it engages focused energy. That energy can easily turn to *distress*, however, if something causes you to view the situation as unmanageable or out of control. Many people regard public speaking or airplane flights as very stressful—causing physical reactions such as an increased heart rate and a loss of appetite—while others look forward to the event. It's often a question of perception: A positive stressor for one person can be a negative stressor for another.

stress

physical and psychological responses to the pressures of daily life

Causes of Stress

The most frequent reasons for "stressing out" fall into three main categories:

1. the unsettling effects of change
2. the feeling that an outside force is challenging or threatening you
3. the feeling that you have lost personal control.

Life events such as marriage, changing jobs, divorce, or the death of a relative or friend are the most common causes of stress. Although life-threatening events are less common, they can be the most physiologically and psychologically acute. They are usually associated with public service career fields in which people experience intense stress levels because of imminent danger and a high degree of uncertainty—police officer, fire and rescue worker, emergency relief worker, and the military.

You may not plan to enter a high-stress career, but as a college student, you may find that the demands of college life can create stressful situations. The National Institute of Mental Health (NIMH) notes some of the more common stressors for college students:

- Increased academic demands
- Being on your own in a new environment
- Changes in family relations
- Financial responsibilities
- Changes in your social life
- Exposure to new people, ideas, and temptations
- Awareness of your sexual identity and orientation
- Preparing for life after graduation (National Institute of Mental Health, 2004).

Symptoms of Distress

Symptoms of stress fall into three general, but interrelated, categories—physical, mental, and emotional. Review this list carefully. If you find yourself frequently experiencing these symptoms, you are likely feeling *distressed*:

- Headaches
- Fatigue
- Gastrointestinal problems
- Hypertension (high blood pressure)
- Heart problems, such as palpitations
- Inability to focus/lack of concentration
- Sleep disturbances, whether it's sleeping too much or an inability to sleep
- Sweating palms/shaking hands
- Anxiety
- Sexual problems

Even when you don't realize it, stress can cause or contribute to serious physical disorders. It increases hormones such as adrenaline and corticosterone, which affect your metabolism, immune reactions, and other stress responses. That can lead to increases in your heart rate, respiration, blood pressure, and physical demands on your internal organs.

Behavioral changes are also expressions of stress. They can include:

- Irritability
- Disruptive eating patterns (overeating or under eating)
- Harsh treatment of others
- Increased smoking or alcohol consumption

Stress management is key to academic success.

- Isolation
- Compulsive shopping

A sustained high level of stress is no laughing matter. It can affect every area of your life—productivity in the workplace and classroom, increased health risks, and relationships, to name just a few.

Managing Stress

As noted in the Introduction, you can learn to manage stress. The first step is understanding yourself better—how you react in different situations, what causes you stress, and how you behave when you feel stressed. Once you've done that, take the following steps:

Set priorities. Use the time-management tips you learned in Section 1. Make a To-Do list. Decide what is really important to get done today, and what can wait. This helps you to know that you are working on your most immediate priorities, and you don't have the stress of trying to remember what you should be doing.

Practice facing stressful moments. Think about the event or situation you expect to face and rehearse your reactions. Find ways to practice dealing with the challenge. If you know that speaking in front of a group frightens you, practice doing it, perhaps with a trusted friend or fellow student. If the pressure of taking tests causes you to freeze up, buy some practice tests at the school bookstore or online and work with them when there are no time pressures.

Examine your expectations. Try to set realistic goals. It's good to push yourself to achieve, but make sure your expectations are realistic. Watch out for perfectionism. Be satisfied with doing the best you can. Nobody's perfect—not you, not your fellow cadet, nobody. Allow people the liberty to make mistakes, and remember that mistakes can be a good teacher.

Live a healthy lifestyle. Get plenty of exercise. Eat healthy foods. Allow time for rest and relaxation. Find a relaxation technique that works for you—prayer, yoga, meditation, or breathing exercises. Look for the humor in life, and enjoy yourself.

Learn to accept change as a part of life. Nothing stays the same. Develop a support system of friends and relatives you can talk to when needed. Believe in yourself and your potential. Remember that many people from disadvantaged backgrounds have gone on to enjoy great success in life.

At the same time, avoid those activities that promise release from stress while actually adding to it. Drinking alcohol (despite what all those TV commercials imply), drinking caffeine, smoking, using narcotics (including marijuana), and overeating all add to the body's stress in addition to their other harmful effects.

Here are some other strategies for dealing with stress:

- Schedule time for vacation, breaks in your routine, hobbies, and fun activities.
- Try to arrange for uninterrupted time to accomplish tasks that need your concentration. Arrange some leisure time during which you can do things that you really enjoy.
- Avoid scheduling too many appointments, meetings, and classes back-to-back. Allow breaks to catch your breath. Take a few slow, deep breaths whenever you feel stressed. Breathe from the abdomen and, as you exhale, silently say to yourself, "I feel calm."
- Become an expert at managing your time. Read books, view videos, and attend seminars on time management. Once you cut down on time wasters, you'll find more time to recharge yourself.
- Learn to say "no." Setting limits can minimize stress. Spend time on *your* main responsibilities and priorities rather than allowing other people's priorities or needs to dictate how you spend your time.
- Exercise regularly to reduce muscle tension and promote a sense of well-being.
- Tap into your support network. Family, friends, and social groups can help when dealing with stressful events (Ayala, 2002).

Depression

depression

a common mental illness that affects a person's body, mood, and thought—it causes people to lose pleasure from daily life, can complicate other medical conditions, and can even lead to suicide

Unfortunately, a person's inability to deal with stress can often lead to clinical **depression**. People with depression have similar symptoms to stress, except the symptoms are not temporary—they can last for weeks at a time. Because of the sustained symptoms, the effect on the body, mood, and behavior is often more serious than with temporary stress. Depression can have severe effects on your eating habits, your relationships, your ability to work and study, and how you think and feel. The illness is not unique to a particular group of people or area of the country. Millions of adult Americans, including many college students, suffer from clinical depression.

It's important to understand that clinical depression is a real, not an "imaginary" illness. It's not a passing mood or a sign of personal weakness. It demands treatment—and 80 percent of those treated begin to feel better in just a few weeks.

According to NIMH, the following symptoms are signs of major depression:

- Sadness, anxiety, or "empty" feelings
- Decreased energy, fatigue, being "slowed down"
- Loss of interest or pleasure in usual activities
- Sleep disturbances (insomnia, oversleeping, or waking much earlier than usual)
- Appetite and weight changes (either loss or gain)
- Feelings of hopelessness, guilt, and worthlessness
- Thoughts of death or suicide, or suicide attempts
- Difficulty concentrating, making decisions, or remembering
- Irritability or excessive crying
- Chronic aches and pains not explained by another physical condition.

It's normal to have some signs of depression some of the time. But the NIMH says that if someone has five or more symptoms for two weeks or longer, or suffers noticeable changes in normal functioning, that person should go to a mental health professional for evaluation. Depressed people often may not be thinking clearly and may therefore not seek help on their own. They frequently require encouragement from others—they "need help to get help."

Mental health professionals say depression among college students is a serious problem. A recent UCLA survey of college freshmen indicates that today's students are feeling more overwhelmed and stressed than students did 15 years ago. The National Mental Health Association reports that more than 30 percent of college freshmen report feeling overwhelmed a great deal of the time (National Mental Health Association, 2005).

If think you might be depressed, you should talk with a qualified health care or mental health professional. The resident advisor in your dorm, the student health center, your family health-care provider, or a clergy member can help steer you to treatment resources. Several effective treatments for depression are available, and—depending on the severity of the symptoms—can provide relief in just a few weeks. But individuals respond differently to treatment. If you don't start feeling better after a few weeks, talk to your treatment provider about other treatments, or seek a second opinion (National Institute of Mental Health, 2004).

Suicide

As noted above, severe depression often manifests itself in thoughts about death or suicide, or in suicide attempts. Many people are understandably uncomfortable talking about suicide, but doing so can save lives. The NIMH reports that in 2000, suicide was the 11th leading cause of death for all Americans and the third leading cause of death for those aged 15 to 24. While women are three times as likely to attempt suicide as men, men are four times as likely as women to succeed (National Institute for Mental Health, 2003).

There are many common myths about suicide:

- *If someone wants to die, nobody can stop that person*. False. Most people thinking about suicide don't want to die: They want help.
- *If I ask someone about suicide, I'll give that person the idea*. False. That you cared enough to ask may offer comfort to the person.
- *Suicide comes "out of the blue."* False. Usually, the person exhibits several warning signs (Vitt & Calohan, 2002).

Critical Thinking

What are some of the stressors you currently face? Develop an action plan to improve your stress management skills by either eliminating a cause of stress or reducing its effects on you. Incorporate at least three techniques described in this section of your textbook.

You should always take suicidal thoughts, impulses, or behavior seriously. If you are thinking or talking about hurting or killing yourself, or know someone who is, *seek help immediately*. The NIMH recommends you turn to your student health center; a family physician; a professor, coach, or adviser; a member of the clergy; a local suicide or emergency hotline (one number is 1-800-SUICIDE); or a hospital emergency room. If you have to, call 911.

Some of the warning signs of suicide include:

- Talking about suicide
- Statements about hopelessness, helplessness, or worthlessness
- Preoccupation with death
- Becoming suddenly happier or calmer
- Losing interest in things one cares about
- Setting one's affairs in order for no apparent reason—such as giving away prized possessions or making final arrangements regarding finances and property.

CONCLUSION

Stress can have consequences far beyond temporary feelings of pressure. While you can't avoid stress, you can learn to manage it and develop skills to cope with the events or situations you find stressful. By learning to cope with stress, and by recognizing the symptoms of depression and the warning signs of suicide, you'll be better prepared to help not only yourself, but also friends, fellow students, and the Soldiers you will someday lead.

Learning Assessment

1. Define stress and list some of the symptoms.

2. Explain what causes stress and list some of the ways to deal with it.

3. What is the difference between stress and depression?

4. List some warning signs of suicide.

Key Words

stress
depression

References

3rd Brigade, 25 Infantry Division (L). May 2005. *Operation Enduring Freedom: Afghan Leader Book 5 April 2004–May 2005.* Retrieved 5 August 2005 from http://rotc.blackboard.com/courses/1/CCR/content/_488714_1/Operation_Enduring _Freedon_Leader_Book_Apr_04___May_05.pdf

Ayala, S. (October–November 2002). Stress. *Health Tips from Army Medicine.* Madigan Army Medical Center, Fort Lewis, WA. Retrieved 2 August 2005 from http://www.armymedicine.army.mil/hc/healthtips/08/stress.cfm

National Institute of Mental Health. (2003). *In Harm's Way: Suicide in America.* Retrieved 31 August 2005 from http://www.nimh.nih.gov/publicat/harmsway.cfm

National Institute of Mental Health. (2004). What do these students have in common? Retrieved 10 August 2005 from www.nimh.nih.gov/publicat/students.cfm

National Mental Health Association. (2005). Finding Hope and Help: College Student and Depression Pilot Initiative. Retrieved 11 August 2005 from http://www.nmha.org/camh/college/index.cfm

Pawelek, J. & Jeanise, S. (March 2004). Mental Health Myths. *Health Tips from Army Medicine.* Retrieved 11 August 2005 from http://www.armymedicine.army.mil/hc/healthtips/13/200403mhmyths.cfm

Vitt, A. & Calohan, J. (April–May 2002). Suicide Warning Signs. *Health Tips from Army Medicine.* Retrieved 11 August 2005 from http://www.armymedicine.army.mil/hc/healthtips/08/suicidewarning.cfm

Section 4

GOAL SETTING AND PERSONAL MISSION STATEMENT

Key Points

1 Defining a Vision

2 Writing a Personal Mission Statement

3 Writing SMART Goals to Support a Vision and Mission

> If you do not know where you are going, every road will get you nowhere.
>
> Henry Kissenger, former US Secretary of State

Introduction

Successful Army leaders, such as George Washington, Dwight D. Eisenhower, and Colin Powell did not attain greatness by luck. They knew what they believed and where they wanted to go. They were men with solid values and defined goals, as well as a clear vision and sense of mission. Because of their resolve, when they were tested, they were ready to meet the challenges they faced.

As a college student and ROTC cadet, you face many decisions in the next few years that will affect the course of your life. This section will discuss how you can develop a **vision** for your life based on your values, write a personal **mission** statement, and set concrete, attainable personal **goals**. GEN Colin Powell noted the value of vision when he wrote down his 13 "rules for command."

GEN Colin Powell's Rules for Command

1. It ain't as bad as you think. It will look better in the morning.

2. Get mad, then get over it.

3. Avoid having your ego so close to your position that when your position falls, your ego goes with it.

4. It can be done!

5. Be careful what you choose. You may get it.

6. Don't let adverse facts stand in the way of a good decision.

7. You can't make someone else's choices. You shouldn't let someone else make yours.

8. Check small things.

9. Share credit.

10. Remain calm. Be kind.

11. Have a vision. Be demanding.

12. Don't take counsel of your fears or nay sayers.

13. Perpetual optimism is a force multiplier (Powell, 1995).

Critical Thinking

Based on GEN Powell's rules, what do you think he values? How could guidelines like these help you at college?

Defining a Vision

vision

a vivid description of the future that focuses your efforts—your vision is a reference point to guide your decisions, planning, and actions for the future

What Words Best Express Your Values?

Ambitious
Adaptable
Career-minded
Capable
Courageous
Family-focused
Flexible
Honest
Independent
Logical
Obedient
Responsible
Self-controlled
Health-focused
Environmentally
 conscious
Spiritually focused
Others?

Your vision is the guiding theme of your personal life and professional career. Your vision is a long-term picture that establishes your priorities for making short-term decisions. Your vision grows out of your values and, as a cadet, from the core Army values you've already studied in previous sections.

Some people ask: "Why have a vision? Why does it matter?" The answer is that vision provides direction for your life and context for your decisions. If you don't know where you are going—any road can take you there. To lead people—whether fellow cadets, fellow students, Soldiers, or employees—you need to know where you want to go. Vision motivates people to perform to their potential and beyond. A vision reduces the likelihood of complacency, drifting, and mediocre performance.

Your strong vision also makes you a role model. When your subordinates see that you have a clearly defined vision in your life and career, they will imitate you. When you become a second lieutenant, your Soldiers will look to you and your vision to provide the framework and context not only for your command decisions but also for the orders you give and the work you ask them to perform.

Many people and organizations find it helpful to record their vision as a vision statement. This forces you to ask yourself some profound questions: What should your vision statement say? What do you want to be known for? What are the most important things in your life? What do you want to achieve? Think also about the model you will want to project to your fellow students, cadets, and, when you become an officer, to your Soldiers. Your vision should include what matters most to you, so when you write your vision statement, you need to reflect on and clarify your values. Think about the words that best express who you want to be.

A vision statement is usually very concise—no more than a sentence or two. So you'll need to select only about five or six values or characteristics to go into your vision statement (see the marginal text for a partial list).

Of course, your priorities will shift as you get older, gain experience, and advance in your career. That's why you should think about your vision statement as a living document, something that can and will change with time. You are not carving your vision in stone. Plan to revisit your statement, reflect on your priorities, and make changes every year or two.

Here's an example of a possible vision statement: "*I am determined to be the best ROTC cadet among my peers, the best student to my teachers, the best athlete to my coaches, and the best son (or daughter) to my parents that I can possibly be. I pledge to use my strengths to better my weaknesses in all areas of my life.*"

Sounds like a promise, doesn't it? That's what a vision statement is: a commitment to your future, put into words to help guide your efforts. Can you draft your vision statement now?

Writing a Personal Mission Statement

mission

your purpose—personal and professional—which guides your planning and actions as you put your vision into practice

The next step is to develop your mission statement. A mission statement describes your fundamental purpose. It guides the planning and implementation of your vision. It's a description that encompasses your own personal objectives, long-term goals, and guiding philosophy—all of which touch your professional life, as well.

In a corporate environment, a mission statement is a description of what an organization wants to accomplish in business. Similarly, your own mission statement should embrace your personal and professional goals. And the best goals come from what *motivates* us.

Once you have reflected on your values and your motivations, you should be able to craft your mission statement without too much trouble. Remember, a mission statement should not be the Ten Commandments for the rest of your life. It should project perhaps three to five years into the future. Just as you will do with your vision, you should revisit your mission statement and adjust it as your life circumstances change—because they *will* change.

Here's an example of a possible mission statement that supports the vision statement above: "*During the next four years in college I want to achieve excellent grades (B+ or higher) in all my coursework. I will also seek experience in a leadership role in a club, team, or activity, and I will actively seek internships, networking opportunities, and other hands-on experience. In addition, I will volunteer in at least one community service organization on a regular basis. I will keep in close touch with my family as often as possible and help my parents with my younger siblings.*"

Notice anything? The mission statement takes the aims of the vision statement and makes them more concrete. The tone of the mission statement is confident and determined. Do any of the aims expressed in this mission statement seem unreasonable or unreachable?

Now, you try it!

Writing SMART Goals to Support a Vision and Mission

To bring your vision to life and accomplish your personal mission, you need to do one more thing: set some definite goals. One way to think about these goals is to think of your vision and mission as your life *strategy* and your goals as those *tactics* that will help you work within and toward that strategy.

Think of goals as the dots you connect to create the picture described by your vision and mission. They are the *short-term milestones* that will keep you on track and help you achieve your greater mission.

But how can you write effective goals? One technique is to write SMART goals, which you read about in the section on Time Management. SMART goals have built-in features that help you attain them. They are specific, measurable, attainable, realistic, and time bound. Here's an example of how to write a SMART goal.

If your vision, as stated above, is to be an excellent student and your mission in support of that is to attain excellent grades in all four years of college, how exactly do you get there? A SMART goal that implements the vision and mission statements you have read above might be to *get an A in math this semester by joining and regularly attending a study group by the third week of the term.* Notice how the goal sets a specific measurable benchmark and an attainable deadline. You can write goals for each of your classes and activities, if you find that helpful.

What Motivates You?

Money
Recognition
Desire to please
Self-satisfaction
Self-worth
Sense of accomplishment
Fear of failure
Physical needs
Faith
Others?

goals

things you set out to do or achieve—short or long term—to fulfill your mission and realize your vision

SMART Goals:

Specific
Measurable
Attainable
Realistic
Time bound

Critical Thinking

What is the difference between a vision statement and a mission statement? Why are both important?

The last step in this process is to prioritize your SMART goals. Put them in order of importance, time due or time required to complete, overall attainability, cost, geographic location, outside help required, or other organizing scheme. You should not randomly list your goals; otherwise, you'll waste time and effort. Whether you do the small stuff first and the challenging ones later—or vice versa—is up to you, but *organize* your time and effort.

Leadership is not magnetic personality—that can just as well be a glib tongue. It is not "making friends and influencing people"—that is flattery. Leadership is lifting a person's vision to higher sights, the raising of a person's performance to a higher standard, the building of a personality beyond its normal limitations.

Peter Drucker

CONCLUSION

Crafting a personal vision statement, mission statement, and SMART goals is a key step in developing your identity as an adult, a college student, an ROTC cadet, and a future Army leader. Vision, mission, and goals will help bring out the best qualities of your personality and make you a desirable role model for your peers and your subordinates in the future.

You never know the limit of your potential until you reach it—and then step farther. A clear vision, a specific mission, and definite goals are important tools that will help you reach and then exceed your expectations.

Learning Assessment

1. Define vision and describe a vision statement.
2. Describe a personal mission statement and its purpose.
3. Describe SMART goals and how they support a vision and mission.

Key Words

vision
goals
mission

References

Powell, C. (with Persico, J.). (1995). *My American Journey*. New York: Random House, Inc.

Department of the Army. (1984). *Quotes for the Military Writer.* (Cited in DA PAM 600-65, Leadership Statements and Quotes. 1 November 1985. Retrieved 12 July 2005 from http://www.army.mil/usapa/epubs/pdf/p600_65.pdf

Section 5

INTRODUCTION TO EFFECTIVE ARMY COMMUNICATION

Key Points

1 The Communication Process

2 Five Tips for Effective Communication

3 Four Tips for Effective Writing

4 Three Tips for Effective Speaking

… an order that *can* be misunderstood *will* be misunderstood.

Field Marshal Helmuth von Moltke

Introduction

Your success as a military leader depends on your ability to think critically and creatively and to communicate your intention and decision to others. The ability to communicate clearly—to get your intent and ideas across so that others understand your message and act on it—is one of the primary qualities of leadership.

While you are a college student, your channels of communication include presentations and term papers. When you become an Army officer, these channels will expand to include training meetings, briefings, and operations orders. As you will see, the means to effective Army communication is to develop your speaking and writing skills so that you can deliver any message to any audience effectively. Keep in mind that communication also includes receiving messages from others through reading and listening.

Early in your Army career, much of your communication is *direct*. For example, coaching your Soldiers often requires communication that is one-on-one, immediate, and spoken. Later in your Army career, as your leadership responsibilities increase, you will inform subordinates and leaders through written orders, procedures, memos, and e-mail. This form of communication is *indirect*—it goes through other people or processes, is time-delayed, and written.

Your ability to communicate—to write, speak, and listen—affects your ability to inform, teach, coach, and motivate those around you. The good news is that you can develop these essential skills. This section will discuss the communication process and then provide you tips for effective writing and speaking.

It is difficult to overemphasize the importance of these skills. In military operations, as elsewhere, the inability to write and speak well can have tremendous costs. History is replete with examples of misunderstood messages. For example, many Civil War scholars believe that victory at Gettysburg may have depended how a subordinate interpreted Confederate GEN Robert E. Lee's use of the word "practicable."

GEN Robert E. Lee

Day One at Gettysburg: Vague Orders Have Significant Consequences

[On the first day of the battle of Gettysburg, Pa., Confederate attacks drove Union troops through the town to the top of Cemetery Hill, a half-mile south.] The battle so far appeared to be another great Confederate victory.

But Lee could see that so long as the enemy held the high ground south of town, the battle was not over. He knew that the rest of the [Union] Army of the Potomac must be hurrying toward Gettysburg; his best chance to clinch the victory was to seize those hills and ridges before they arrived. So Lee gave [LTG Richard S.] Ewell discretionary orders to attack Cemetery Hill "if practicable." Had [LTG Thomas J. (Stonewall)] Jackson still lived, he undoubtedly would have found it practicable. But Ewell was not Jackson. Thinking the enemy position too strong, he did not attack—thereby creating one of the controversial "ifs" of Gettysburg that have echoed down the years (McPherson, 1988).

The Communication Process

sender

the person who originates and sends a message

As you will see, the Gettysburg vignette illustrates the parts of the communications process. Lee was the **sender**. He sent the message: Attack Cemetery Hill. Ewell, the **receiver**, read the words "if practicable," decided that Union artillery on the hill made an attack not "practicable," and did not attack.

receiver

the person who receives the sender's message, or for whom the sender intends it

The words "if practicable" made the message vague. (Who and what should define "if practicable?") Obstacles to communication, such as this lack of clarity—along with other considerations, such as the demands of time, the ease of understanding the sender's speech, the ability to read the sender's handwriting, or the distractions in the area—make up what communications theorists call **noise**. Noise works against the clarity of communication.

Looking again at the vignette above, you find that Lee never checked with Ewell to see if he understood Lee's intent: "What do you intend to do?" Ewell never checked with Lee to clarify the message: "What do you mean by 'if practicable'?" The communication process included no *feedback*. Assume for a moment, as some historians do, that Lee intended that Ewell attack Cemetery Hill immediately and decisively. (These historians argue that Lee was used to issuing such vague orders to the aggressive Stonewall Jackson, who had died a few months earlier.) Throwing Billy Yank off the hilltop might well have allowed Lee to command the battlefield, perhaps even forcing the advancing Union armies to withdraw. That might have led to Lee's domination of southern Pennsylvania, choking off Washington from the North and ending the war on the Confederacy's terms.

noise

whatever interferes with communication between the sender and receiver, from the wording used to audience distractions to bad handwriting

If that *was* Lee's intent, the message failed. Ewell did not attack. The Union held onto the high ground and won the battle two days later—the beginning of the end for the Confederacy.

There's an important lesson is all this: *Effective communication occurs when the receiver's perceived idea matches the sender's intended idea.* The receiver understands what the sender *means*, not just what the sender says or writes. But how do you ensure that occurs?

Five Tips for Effective Communication

These five tips will help you eliminate noise and ensure that your receiver understands your message.

1. Focus your message

Every academic, business, or military message you will ever produce will fit into one of two categories:

- *Action-and-information messages* ask the receiver to do something: Schedule a make-up exam; prepare a marketing report; attack a hilltop.
- *Information-only messages* tell the receiver something: The primary cause of the American Civil War was states' rights; Estelle LaMonica is the new Vice President of Human Resources; Alpha Company has one vehicle down for battle damage.

You must focus—*clarify*—your message so your receiver is certain—*clear*—on what he or she is supposed to do or know. Too many action-and-information messages fail because the receiver mistakes them for information-only messages:

"I need a make-up exam," you send. "*You sure do,*" *thinks the receiver.*

"If we knew the market better, we could increase our share." "*That's a good idea.*"

"The bad guys have a company-sized element on Hill 442." "*That's right. I saw the intelligence reports as well.*"

Decide *before you communicate* if your message is action-*and*-information or information-*only*. If you're communicating an action-and-information message, specify what your receiver must *do* and *know*. If you are communicating an information-*only* message, specify what your receiver needs to *know*.

2. Break through the noise

As the sender, as the one trying to communicate, you have the responsibility to communicate clearly—to break through the noise. Think in terms of your receiver. Use your receiver's terms of reference. If a military objective is to your front but to your receiver's flank, refer

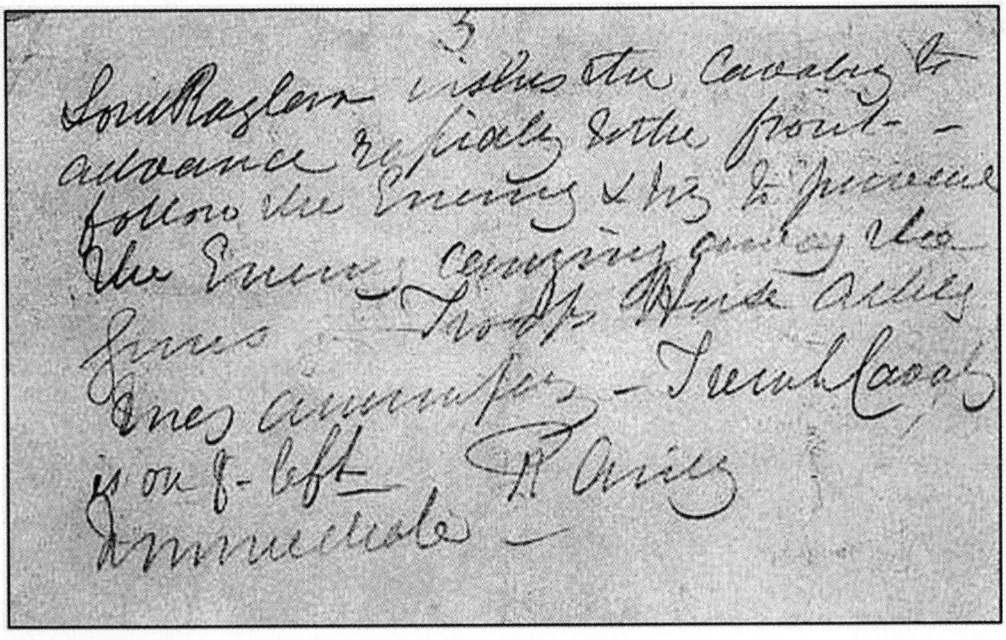

Figure 5.1 The Battle of Balaklava, 25 October 1854—Lord Raglan's reference to his front, rather than the cavalry's flank, sent the Light Brigade to its death: "Lord Raglan wishes the cavalry to advance rapidly to the front, follow the enemy and try to prevent the enemy carrying away the guns. Troop Horse Artillery may accompany. French cavalry is on your left. Immediate."

Critical Thinking

What action requests in the messages under "Focus Your Message" on page 79 were lost in transmission?

to the objective as to the flank. (This very mistake—*front* rather than *flank*—resulted in the deaths of 550 British cavalry troops at the Battle of Balaklava in the Crimean War.)

- **Use descriptive language**. Use visualization and analogies. Instead of saying, "The motor pool is big," say, "The motor pool is the size of a football field." Instead of saying, "I want you to snap that salute," say, "I want you to snap that salute like you were saluting a Normandy veteran."
- **Ask for feedback**. It is not enough to ask, "Do you understand me?" The obvious answer is "Yes. Absolutely. Sure I do," *no matter what the understanding may be.* Lee's intent may have been "Take Cemetery Hill." Ewell's understanding may have been "Take Cemetery Hill only if I can do it without casualties." If Lee had asked, "Do you understand me?" Ewell's response—no matter the difference between intent and understanding—would have been "Yes. Absolutely. Sure I do." Craft your request for feedback so your receiver will have to demonstrate his or her understanding of the message. "What—specifically—do I want you to do?" "What—specifically— will you do now?" "How will you do it?"
- **Revise as you need to**. You may have to repeat your message several times before you communicate. Use the feedback you get to adjust your message to the needs of your receiver.

3. Put your Bottom Line Up Front (BLUF)

Get to your point in the first 10 seconds of your message; put your **bottom line up front (BLUF)**.

- Your point—your *bottom line*—in an action-and-information message is what you want your receiver to do: "Attack, seize, and hold Cemetery Hill."
- Your point—your bottom line—in an information-only message is what you want your receiver to know: "Alpha Company has one vehicle down for battle damage."
- Audiences—receivers—are impatient. "Get to the point," they say. "How does this affect me?" If you don't get to your point, if you don't explain how your message affects your receivers, they will tune out. They may be physically present during the rest of the message, but their minds are far away—thinking of food; thinking of home; thinking of other tasks and responsibilities they have to perform.

4. Use simple words

Great communicators use simple words.
  Consider these examples:

  "Carthage must be destroyed!" (Cato the Elder, an ancient Roman senator)
  "The only thing we have to fear is fear itself." (Franklin D. Roosevelt)
  "I have a dream." (Dr. Martin Luther King, Jr.)
  "Mr. Gorbachev, tear down this wall!" (Ronald Reagan)

Look closely at the examples. Notice the overwhelming use of single-syllable words. (Of the 24 words, only five are two-syllable words.) Notice the absence of any long "impressive" words. Given the choice between a simple word and a long word—and given there's no difference in the meaning of the two words—use the simple word. Your communication will be clearer.

5. Use concrete words

Concrete words draw pictures in your receiver's brain.
Consider the difference between these two phrases:

"An old car."
"A 1966 red Mustang convertible."

Which phrase draws a picture in your brain? You can visualize the Mustang far more easily, far more quickly, than you can visualize an old car. You can visualize "15 enemy soldiers with small arms and shoulder-fired anti-tank weapons" far more quickly, far more easily, than you can visualize "a bunch of bad guys."

Four Tips for Effective Writing

Writing takes special care. You can reread and study a written message, while a spoken message quickly vanishes into the air. As you saw above, an unclear written message can lead to disaster, especially if the receiver has no way to confirm his or her understanding of the message. These tips will help ensure your writing is as clear as possible.

1. Use the "Five Tips for Effective Communication"

The five tips—focus your message, break through the noise, put your bottom line up front, use simple words, and use concrete words—will make you a better writer. Because you don't have an audience in front of you and because you have no immediate feedback, clarity becomes critical. You must not only be sure the receiver understands you; but you must also remove the opportunity to misunderstand.

2. Use active voice, short sentences, and conversational language

Use active voice. *Active voice* describes a sentence in which the subject of the sentence performs the action of the sentence: "Sergeant Torres wrote the report." "Sergeant Torres" is the subject of the sentence; he is what the sentence is about. He does the action of the sentence: He writes the report.

Passive voice—the less-effective counterpart of active voice—describes a sentence in which the subject of the sentence receives the action: "The report was written by Sergeant Torres." Now the subject is "The report"; it receives the action; it "was written."
Active voice has three advantages over passive voice:

1. *It's more concise.* "Sergeant Torres wrote the report" has five words. "The report was written by Sergeant Torres" has seven. Active voice will *almost always* be more concise than passive voice.
2. *It's more direct.* It demands accountability. You cannot write in active voice unless you identify the doer of the action. Consider the ethical implications of "No action was taken." Who didn't take action? Why didn't they take action? And why didn't the writer name whoever didn't take action?
3. *It's more conversational.* It's more natural. You grow up speaking in active voice. When you were little, you may have said, "I want to be a soldier." You certainly didn't say, "To be a soldier is wanted by me."

A classic and valuable writing guide is The Elements of Style, by William Strunk Jr. and E.B. White. The Fourth Edition is available online and in many bookstores. You can also find useful information and tips at Purdue University's Online Writing Lab (OWL), www.english.purdue.edu.

active voice

in the active voice, the doer of the action is the subject of the verb

passive voice

in the passive voice, the subject of the verb receives the action— avoid this weak construction

> ## A document that looks hard to read is hard to read.
> Diane Brewster-Norman, communications-skills expert

Use short sentences. Short sentences are easier to read. They are easier to keep grammatically correct, and they are easier to punctuate. Keep your sentences to an average of 12 to 15 words per sentence.

Use conversational language. Use the language you use every day. As you are writing, ask yourself, "How would I say this?" In conversational language, you would never say "Upon completion of the above-entitled actions, forward the documents to the undersigned." You would probably say, "When you are done with this, return the papers to me." There is, however, a caution. Conversational written language does not exactly match the spoken language. It doesn't include the "um"s and "uh"s. It doesn't include the half-sentences people start, then change.

3. Use lots of white space

White space lets your reader breathe. Keep your paragraphs to no more than about six lines long. Use headings and lists. Open up your document.

Examine this textbook section. Notice the short paragraphs, the headings, and the lists. It should look easy to read. The writers wrote it that way. Make your documents easy to read.

4. Use correct grammar, spelling, and punctuation

Using incorrect grammar, spelling, and punctuation presents two problems.

It can be confusing. What happens when you read, "We saw a motor pool walking through the battalion area?" You are not sure what the writer intended: Motor pools don't generally walk, let alone through battalion areas. Try "Walking through the battalion area, we saw a motor pool."

It affects your credibility. Readers assume that if you cannot take care of the little things, you cannot take care of the big things. If you cannot write a simple sentence, they wonder, how can you lead troops? The assumption may not be fair (GEN Ulysses S. Grant was a horrible speller), but it's real and has hampered many officers' careers.

A complete discussion of grammar, spelling, and punctuation is beyond the scope of this lesson, but here are three ways to improve your language ability.

Read professionally written and published material. The subject doesn't matter—as long as it is professionally written and published. Read good books. See the written word on the page. Get used to the standards of written English. As you become familiar with the standards, you will see your mistakes more easily.

Make writing skills a part of your professional development plan. Learn about grammar, spelling, and punctuation. Learn the principles. Learn the forms. Learn the expectations. Then coach your Soldiers.

Get someone to review your work. All too often, you will be too close to your document to see your errors. Your spell-checker won't catch errors that are spelled correctly.

Critical Thinking

Assuming that Lee wanted Ewell to take Cemetery Hill, how might the commander have written his order to make it clearer to Ewell what he intended? If Lee wanted Ewell to be cautious, how might he have written it?

Three Tips for Effective Speaking

Public speaking and briefing also require a mastery of the language, but involve different skills than writing. These tips will help.

1. Use the "Five Tips for Effective Communication"

The five tips—focus your message, break through the noise, put your bottom line up front, use simple words, and use concrete words—will make you a better speaker. Consider that the four examples you read in "Use simple words" (from Cato, Roosevelt, King, and Reagan) were all originally spoken.

2. Mark the parts of your presentation

Look at the page in front of you. Besides the words, you'll see headings, paragraphs, and lists. These mark the parts of the reading. As you move from one paragraph to another, you expect the ideas to transition or flow from one to the next. The page layout reinforces the ideas in the reading. The spoken word provides no such markers, however. There's no white space, no indents, no bolded lists. So you provide the markers.

> **Use pauses to indicate changes in ideas**. When you've finished with an idea, pause for a few seconds. Count the seconds in your head. *One. Two. Three. Four. Five.* Then pick up the conversation. The silence—the pause—represents the white space on a page. You're moving to another point.

> **Use movement to indicate changes in ideas**. If you're standing on your audience's left front while you're explaining your first point, move to your audience's right front to explain your second point.

> "That concludes my first point." (Stop. Step. Step. Step.) "My second point . . . ." The physical movement represents the movement from your first point to your second.

> **Use gestures to indicate the parts of your presentation**. You've worked very carefully to structure your presentation. Use your gestures as body language to complement that structure. "The first part of the five-paragraph field order . . . ." (Hold up your thumb or index finger.) " . . . is 'situation.'"

3. Listen actively

Speaking has certain advantages over writing. When you speak, you have your audience members in front of you. You get immediate feedback from them. You can observe their body language and determine how your message is going over.

Positive signs include audience members learning forward, listening to what you say. They nod their heads in agreement. They make eye contact with you. They look at your slides.

Negative signs include closed body language: Audience members lean back in their chairs and fold their arms across their chests. They glance at their watches. (Some may check to see if their watches are still running.) They look around the room. The negative signs mean you need to change your approach or your delivery. Perhaps the best way is to pause and ask your audience for direct feedback: "I get the impression you're not comfortable with this discussion. What are your concerns?" Better to address the issues than ignore them.

CONCLUSION

Think of the great communicators of the last century: Franklin Roosevelt, Winston Churchill, Margaret Thatcher, Martin Luther King, Jr., and Ronald Reagan. Would these men and women have *led* as well as they did if they didn't *communicate* as well as they did? You will soon lead young men and women. You cannot lead unless you can communicate.

Learning Assessment

1. How many parties does it take for communication to take place? Who are they?
2. Explain what BLUF means and why it is important.
3. How do you know if someone has understood you?
4. What are three reasons it is generally better to use active rather than passive voice?
5. Describe the five tips for effective communication.

Key Words

sender
receiver
noise
feedback
BLUF
active voice
passive voice

References

Department of the Army. PAM 600–67, *Effective Writing for Army Leaders*. 2 June 1986.
McPherson, J.M. (1988). *Battle Cry of Freedom*. New York: Ballantine Books.
ST 22-2, *Writing and Speaking Skills for Leaders at the Organizational Level*. August 1991.

INTRODUCTION TO THE WARRIOR ETHOS

Key Points

1 The Warrior Ethos Defined

2 The Soldier's Creed

3 The Four Tenets of the Warrior Ethos

Every organization has an internal culture and ethos. A true Warrior Ethos must underpin the Army's enduring traditions and values. It must drive a personal commitment to excellence and ethical mission accomplishment to make our Soldiers different from all others in the world. This ethos must be a fundamental characteristic of the U.S. Army as Soldiers imbued with an ethically grounded Warrior Ethos who clearly symbolize the Army's unwavering commitment to the nation we serve. The Army has always embraced this ethos but the demands of Transformation will require a renewed effort to ensure all Soldiers truly understand and embody this Warrior Ethos.

GEN Eric K. Shinseki

Introduction

Every Soldier must know the Soldier's Creed and live the Warrior Ethos. As a cadet and future officer, you must embody high professional standards and reflect American values. The Warrior Ethos demands a commitment on the part of all Soldiers to stand prepared and confident to accomplish their assigned tasks and face all challenges, including enemy resistance—anytime, anywhere.

This is not a simple or easy task. *First*, you must understand how the building blocks of the Warrior Ethos (see Figure 1.1) form a set of professional beliefs and attitudes that shape the American Soldier. *Second*, you must establish an unwavering personal commitment to excellence and ethical mission accomplishment, a commitment that cannot vary, no matter what the circumstances. *Finally*, as a leader, you must be the example for your Soldiers of what it means to live the Warrior Ethos, through your own conduct.

This section defines the Warrior Ethos, covers its four tenets as based on a commitment to selfless service to the nation and the Army values, and demonstrates how the Soldier's Creed ties its concepts together.

The following vignette epitomizes the power of the Warrior Ethos in the Contemporary Operating Environment, a commitment to the welfare of others so strong that it sets a timeless example of sacrifice for one's fellow Soldiers.

MSG Gordon and SFC Shughart in Somalia

During a raid in Mogadishu in October 1993, MSG Gary Gordon and SFC Randall Shughart, leader and member of a sniper team, respectively, with Task Force Ranger in Somalia, were providing precision and suppressive fires from helicopters above two helicopter crash sites. Learning that no ground forces were available to rescue one of the downed aircrews and aware that a growing number

Figure 1.1   The building blocks of the Warrior Ethos.

of enemy were closing in on the site, MSG Gordon and SFC Shughart volunteered to be inserted to protect their critically wounded comrades.

Their initial request was turned down because of the dangerous situation. They asked a second time; permission was denied. Only after their third request were they inserted.

MSG Gordon and SFC Shughart were inserted one hundred meters south of the downed chopper. Armed only with their personal weapons, the two NCOs fought their way to the downed fliers through intense small arms fire, a maze of shanties and shacks, and the enemy converging on the site. After MSG Gordon and SFC Shughart pulled the wounded from the wreckage, they established a perimeter, put themselves in the most dangerous position, and fought off a series of attacks. The two NCOs continued to protect their comrades until they had depleted their ammunition and were themselves fatally wounded. Their actions saved the life of an Army pilot.

No one will ever know what was running through the minds of MSG Gordon and SFC Shughart as they left the comparative safety of their helicopter to go to the aid of the downed aircrew. The two NCOs knew there was no ground rescue force available, and they certainly knew there was no going back to their helicopter. They may have suspected that things would turn out as they did; nonetheless, they did what they believed to be the right thing. They acted based on Army values, which they had clearly made their own: *loyalty* to their fellow Soldiers; the *duty* to stand by them, regardless of the circumstances; the *personal courage* to act, even in the face of great danger; *selfless service*, the willingness to give their all. MSG Gary I. Gordon and SFC Randall D. Shughart lived Army values to the end; they were posthumously awarded Medals of Honor (FM 22-100).

---

ethos

the disposition, character, or fundamental values peculiar to a specific person, people, culture, or movement

Warrior Ethos

the professional attitudes and beliefs that characterize the American Soldier—the Warrior Ethos is the foundation for the American Soldier's total commitment to victory in peace and war

The Warrior Ethos Defined

Ethos is defined as the disposition, character, or fundamental values peculiar to a specific person, people, culture, or movement. The **Warrior Ethos**, the professional attitudes and beliefs that characterize the American Soldier, is a reflection of our nation's enduring values by the profession charged with protecting those values. The Warrior Ethos is the foundation for the American Soldier's total commitment to victory in peace and war.

At the core of every Soldier is the willingness and desire to serve the nation—both its people and its enduring values. Hence, the foundation for the pyramid representing the Warrior Ethos is a commitment to serve the nation. Soldiers who live the Warrior Ethos put the mission first, refuse to accept defeat, never quit, and never leave a fallen comrade. They have absolute faith in themselves and their team, because they have common beliefs and values.

The **seven Army values** of Loyalty, Duty, Respect, Selfless Service, Honor, Integrity, and Personal Courage (LDRSHIP) form the second level of this pyramid of the Warrior Ethos. Army values are universal; they enable us to see what is right or wrong in any situation. When you encounter a situation that requires you to make a decision, you should apply the Army values. If any one term is not applied, the decision will be flawed. As you can see, Army values and the Warrior Ethos are integral parts of a unified system of beliefs—as with the Soldiers who follow them, they depend on each other. The **Soldier's Creed** ties this system together.

creed

a statement of beliefs, or a statement of a belief and a system of principles or opinions

The Soldier's Creed

The Soldier's Creed, first committed to memory and then increasingly applied to all your tasks—whether routine and safe, or urgent and dangerous—puts the Warrior Ethos into the practical context of the Army Leadership Framework: Be, Know, and Do. The intent of the Soldier's Creed is to link your commitment to selfless service to the goal of every other American Soldier—*victory with honor*.

The Soldier's Creed unifies the Army's culture by expressing fundamental human beliefs from a warrior's perspective. It helps Soldiers understand that, no matter what their personal or professional backgrounds may be, all Soldiers are warriors and members of a team with difficult and dangerous tasks to perform. To develop into an effective leader of Soldiers, you must begin now to live by the seven Army values, the Warrior Ethos and the Soldier's Creed.

The Four Tenets of the Warrior Ethos

- Always place the mission first
- Never accept defeat
- Never quit
- Never leave a fallen comrade

While all citizens hold beliefs and values that bring our nation together, Soldiers must take *action* to protect the nation. The four tenets of the Warrior Ethos provide the *motivation* for that action; motivation built on a comradeship that the Warrior Ethos creates. Because of that comradeship, Soldiers fight for each other, as well as for their nation and for their beliefs. Time and again you see that Soldiers would rather die than let their buddies down. It will be your job as a leader to ensure that your unit has the final ingredients necessary for victory. You must train and lead your Soldiers to become a competent, confident, flexible, and adaptable team—a team imbued with the Warrior Ethos.

Just such a team rescued more than 500 American and Allied prisoners of war from the Japanese at the end of World War II.

The Soldier's Creed

I am an American Soldier. I am a warrior and a member of a team. I serve the people of the United States and live the Army values. I will always place the mission first. I will never accept defeat. I will never quit. I will never leave a fallen comrade. I am disciplined, physically and mentally tough, trained and proficient in my warrior tasks and drills. I always maintain my arms, my equipment and myself. I am an expert and I am a professional. I stand ready to deploy, engage, and destroy the enemies of the United States of America in close combat. I am a guardian of freedom and the American way of life. I am an American Soldier.

Great Raid on Cabanatuan Depicts Warrior Ethos

WASHINGTON (Army News Service, Aug. 10, 2005)—It was one of the most daring and successful Special Operations missions of World War II, full of drama, suspense and heroism—just the sort of thing that would make an exciting movie.

The 1945 raid by the U.S. Army's 6th Ranger Battalion to rescue Americans held at the Japanese POW camp near Cabanatuan in the Philippines is the subject of [the movie] "The Great Raid. . ."

The same raid was depicted in the opening scenes of an earlier movie, the 1945 "Back to Bataan," starring John Wayne and Anthony Quinn.

Regardless of how accurately either movie depicts the raid and those who lived through it, the real-life story is one worthy of study. It is noteworthy as an example of a well-planned and expertly conducted small-unit mission.

It may be even more valuable, however, as a reminder that the Warrior Ethos and Soldiers Creed that American Soldiers live by today are neither new nor exclusive to the men and women on the front lines in Iraq, Afghanistan, and elsewhere around the world.

Great Soldiers of the past lived and fought by those values. There are few better examples of this than what was done by the men of the 6th Ranger Battalion answering the call to duty in late January 1945.

'I will always place the mission first'

The more than 500 Americans inside the barbed wire of the Cabanatuan POW camp in early 1945 were survivors from America's darkest days, the fall of the Philippines in 1942. They were the lucky ones—if "lucky" means staying alive to be continually starved and mistreated by their captors.

Somehow these Soldiers, Marines, Sailors and Airmen, as well as American civilians and some allies, had survived the valiant but doomed battles of Bataan and Corregidor. Somehow many of them had survived the Bataan Death March, which followed Bataan's surrender on April 9, 1942 (Corregidor surrendered on May 6).

'I will never quit'

Somehow they had survived almost three years of starvation, mistreatment, minimal medical care and executions for various offenses proscribed by their guards. Somehow, they had missed the fate of thousands of their comrades who had died when American planes and submarines attacked and sank Japanese ships transporting them from the Philippines. The ships bore no indication of the human cargo they were carrying, so they were routinely attacked by the U.S. Navy and Army Air Force in the campaign to cut the enemy's supply lines.

As U.S. forces returned to the Philippines on Oct. 20, 1944, with the landing at Leyte, followed on Jan. 9, 1945, by landing on Luzon, the question became whether the POWs would be liberated before time ran out for them. It wasn't only a matter of malnutrition and disease catching up to the prisoners or their being moved farther away from the advancing American forces; it was whether they would be murdered before they could be freed.

This was a very real possibility. About 150 American prisoners at a POW camp on the Philippine island of Palawan had been killed by their guards on Dec. 14, 1944. A survivor of this massacre had reached friendly forces and what had happened was known to U.S. Army intelligence by the time of the Luzon invasion. [The possibility of] a similar fate for any captive Americans on Luzon could not be overlooked.

Rescuers: 'I will never accept defeat'

The U.S. Army was determined those who had upheld America's honor in the opening days of the war would not suffer [such] a fate.

To that end, the commanding general of Sixth U.S. Army, Lt. Gen. Walter Krueger, called on the commander of a unique unit under his command, the 6th Ranger Battalion, the only Ranger battalion in the Pacific theater (During World War II, the Army had six Ranger battalions. The 1st through the 5th fought in either the Mediterranean or European theaters; the 6th fought in the Philippines.)

Lt. Col. Henry A. Mucci, a 1936 graduate of West Point, commanded the 6th Ranger Battalion. He had taken command of it in April 1944 in New Guinea when it was the 98th Field Artillery Battalion and led it through its re-designation and transformation into the 6th Ranger Battalion, putting its members through a demanding training program and weeding out those who couldn't or wouldn't measure up to Ranger standards.

By January 1945, his men were all volunteers and ready for a mission. The 6th Rangers landed on three islands in Leyte Gulf Oct. 17, and performed some commando-type missions. Now they were called upon to raid the Cabanatuan POW camp. Specifically, Mucci was to infiltrate about 30 miles behind enemy lines, reach the camp, overcome the guard force, liberate the prisoners and return them safely to friendly lines before the Japanese could react.

The ground to be covered was open and great care would have to be taken to avoid being spotted enroute to the camp. In addition to overcoming the camp's guard force, there were numerous other enemy forces in the area. Because of its proximity to major roadways, the camp often played host to Japanese units in transit. Due to American aircraft, the Japanese made troop movements at night.

A Japanese battalion regularly bivouacked about a mile from the camp and a division-sized unit was believed to be around Cabanatuan City, three to four miles from the camp. These Japanese units had tanks and tanks were also known to be included in the nocturnal movements around the camp.

To accomplish the mission, which he would personally lead, Mucci chose one company of the 6th Rangers, Company C, commanded by Capt. Robert W. Prince. Company C would be reinforced by the 2nd Platoon of Company F, led by 1st Lt. John F. Murphy. The Ranger force would also include four combat photographers from the 832nd Signal Service Battalion and two teams of Sixth Army's elite recon unit, the Alamo Scouts. Counting a few additions from elsewhere in the battalion, the Ranger force consisted of about 120 men.

The Rangers would receive invaluable support from several hundred Filipino guerrillas under the commands of Captains Eduardo Joson and Juan Pajota. The guerrillas would provide intelligence, carry out security along the route to and from the camp, and interface with the civilian population for needed support for the Rangers and the liberated prisoners. The guerrillas would also play a critical role during the assault on the camp.

'I will never leave a fallen comrade'

When Mucci briefed them on the mission, the Rangers immediately knew just how important it was and how difficult it was going to be to pull it off. Each was given the opportunity to stay back. None took it.

It was clear to all of them that they were the only hope to bring out the survivors of Bataan and Corregidor before the Japanese killed them. Mucci ordered them to take an oath to die fighting before letting any harm come to those they were to rescue.

The Raid

The Rangers moved out early on Jan. 28 and soon linked up with guerillas commanded by Joson. By dark, the combined Ranger-guerilla force was inside enemy territory.

At the village of Balincarin, the Rangers were provided the latest intelligence from the Alamo Scouts, who had started their recon duties a day earlier. They were also joined there by Pajota's guerilla force. Working with Pajota, Prince coordinated for the guerillas to provide security, collect enough carabao carts to transport liberated POWs too weak to walk back and prepare enough food for several hundred men.

Mucci delayed the raid for a day in order to gather additional intelligence and to allow a large force of Japanese transiting the area to move away from the camp. The delay also allowed the Rangers to gather detailed information on the camp and its defenders.

The plan for the night-time assault on the compound gave the two guerilla forces the vital mission of stopping any enemy reaction forces coming from nearby Cabanatuan City and Cabu. A Ranger bazooka section would be attached to the guerillas to deal with expected Japanese tanks. The other Rangers would hit the camp from two sides, with Murphy's 2nd Platoon of Company F assaulting the rear entrance and Prince's Company C storming through the front gate of the camp. To distract the guards while the Rangers positioned themselves for the assault, a P-61 night fighter would fly overhead just prior to the attack.

The Rangers and guerillas moved into position at twilight on Jan. 30. The force attacking the front of the camp had to crawl a mile across open ground to reach their jump-off position. The overflight by the night flyer worked as planned, drawing the attention of both guards and prisoners to the sky.

At 7:45 p.m., Murphy on the rear side of the compound fired the first shot, the signal for the attack to commence. The Rangers hit the Japanese soldiers with overwhelming ferocity, using every weapon they had. They concentrated initially on the guard towers, pillboxes and all Japanese in the open. When all enemy positions had been neutralized, the Rangers stormed into the compound and continued to eliminate enemy soldiers and interior defensive positions.

Meanwhile the guerillas at the blocking positions had their own battle to fight. Pajota's men opened fire on the Japanese battalion in the bivouac next to Cabu Creek. Guerilla machine gunners stopped the Japanese counterattacks at the Cabu Creek bridge while the Ranger bazooka teams knocked out two tanks and a truck.

The other roadblock under Joson was not attacked, thanks to attacks by P-61 night fighters on a Japanese convoy headed toward Joson's position.

In less than 15 minutes, all serious resistance inside the POW compound had been eliminated, though a final trio of mortar rounds wounded six men and mortally wounded the battalion surgeon, one of only two Rangers to die in the attack. A total of seven were injured.

Within half an hour from the opening shot by Murphy, Prince had completed two searches of the camp and had determined all the prisoners had been found and removed from the camp. Although no prisoners were killed during the fighting, one weakened man suffered a fatal heart attack while leaving the camp.

One British POW who hid in the latrines during the raid wasn't found by the Rangers, but he was picked up the next day by Filipino guerrillas.

The Rangers and liberated prisoners made their withdrawal while Pajota continued to stop all Japanese attempts to pursue. By the time Pajota's men disengaged, they had essentially destroyed an enemy battalion while suffering no fatalities or serious wounds themselves.

Filipino citizens provided food and water to the liberated prisoners on the route back. Additional carabao carts arrived to transport former prisoners too weak to walk. The guerillas continued to provide all-around security.

About 12 hours after the assault on the camp, radio contact was made with Sixth Army. Trucks were requested to meet the force. A couple of hours later, the Rangers and prisoners returned to friendly lines and shortly thereafter, the heroes of Bataan and Corregidor were undergoing medical examination at the 92nd Evacuation Hospital.

The mission, which rescued 511 American and Allied POWs and killed or wounded some 520 Japanese at the cost of two Rangers killed, was completed.

The Cabanatuan raid rescuers and rescued may not have been able to recite the Warrior Ethos of today's Army, but they lived it (Pullen, 2005).

Critical Thinking

How can you, as a cadet, begin to live the Warrior Ethos in your ROTC activities and your daily life on campus?

CONCLUSION

The Warrior Ethos is your commitment to overcome all obstacles. It reflects a selfless dedication to the nation, mission, unit, and your fellow Soldiers. You will develop and maintain this attitude through discipline, rigorous training, learning and embodying Army values, and recognizing that as a cadet you represent the future of the Army's proud heritage.

Learning Assessment

1. Recite the Soldier's Creed from memory.
2. List the four building blocks of the Warrior Ethos.
3. Identify the four tenets of the Warrior Ethos.
4. List the seven Army values.

Key Words

ethos
Warrior Ethos
seven Army values
creed
Soldier's Creed

References

Field Manual 22-100, *Army Leadership: Be, Know, Do*. 31 August 1999.

Pullen, R. (10 August 2005). Great Raid on Cabanatuan Depicts Warrior Ethos. *Army News Service*. Retrieved 17 August 2005 from http://www4.army.mil/ocpa/read.php?story_id_key=7723

Shinseki, E. K. (n.d.) Warrior Ethos. *Leaders' Perspective*. TRADOC News Service. Retrieved 14 August 2005 from http://www.tradoc.army.mil/pao/Web_specials/WarriorEthos/leaderpersp.htm

ROTC RANK STRUCTURE

Key Points

1 The Purpose of Army Ranks

2 The Cadet Ranks

3 The Cadet Unit Structure

4 The Cadet Chain of Command

As the Continental Army have unfortunately no uniforms, and consequently many inconveniences must arise from not being able always to distinguish the commissioned officer from the non-commissioned, and the noncommissioned from the privates, it is desired that some badges of distinction may be immediately provided. For instance, that the field officers may have red or pink-colored cockades in their hats, the captains yellow or buff and the subalterns green. They are to furnish themselves accordingly. The sergeants may be distinguished by an epaulette or stripe of red cloth sewed upon their right shoulder; the corporals by one of green (Long, 1895).

GEN George Washington

Introduction

Your **rank** shows where you fit in the **chain of command**, and the chain of command provides the leadership structure for military **units**.

As a new cadet, you are responsible for following the directions, guidance, and example of those who outrank you. As you advance through ROTC, you will have the opportunity to lead progressively larger and more complex organizations, from the smallest—the team—through the largest—the ROTC battalion.

Military rank is a critical part of the profession of arms. The Continental Army—the army that won independence from Great Britain—at first had no uniforms or badges of rank. The army was made of farmers, laborers, and shopkeepers who wore their work clothes to drill and battle. Think of the confusion that an army without uniforms and rank might experience.

The Purpose of Army Ranks

Military ranks identify who is in charge, indicate levels of leadership and responsibility, and support fast and effective decision making and problem solving.

The Continental Army drew its rank structure—lieutenant, captain, major, colonel, and general—from the traditions of Great Britain, and today's structure remains close to that of the Continental Army. Lieutenants and captains are "company grade" officers, majors and colonels are "field grade" officers, and general officers are "flag" officers.

The Cadet Ranks

ROTC has six cadet officer ranks. The ranks themselves come from British tradition, and the insignia resemble British, European, and colonial insignia.

Cadet second lieutenants—the most junior of the officer ranks—wear a single disc or dot. Cadet first lieutenants wear two discs or dots. Cadet captains wear three discs or dots.

The insignia changes at major—the first of the field-grade officers. Cadet majors wear a single diamond (sometimes called a *lozenge*), cadet lieutenant colonels wear two diamonds, and cadet colonels wear three diamonds.

ROTC recognizes eight cadet noncommissioned officer (NCO) ranks. The ranks and the insignia resemble those of the active Army. Cadet corporals—the most junior of the NCO ranks—wear two chevrons (sometimes called *stripes*). Cadet sergeants wear three

 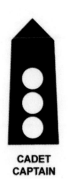   

| CADET SECOND LIEUTENANT | CADET FIRST LIEUTENANT | CADET CAPTAIN | CADET MAJOR | CADET LIEUTENANT COLONEL | CADET COLONEL |

Figure 1.1   Cadet Officer Ranks

Note that the plural form of sergeant major is **sergeants major**.

Abbreviations for Army
Enlisted Ranks
PVT—private
PFC—private first class
SPC—specialist
CPL—corporal
SGT—sergeant
SSG—staff sergeant
SFC—sergeant first class
MSG—master sergeant
1SG—first sergeant
SGM—sergeant major
CSM—command
    sergeant major

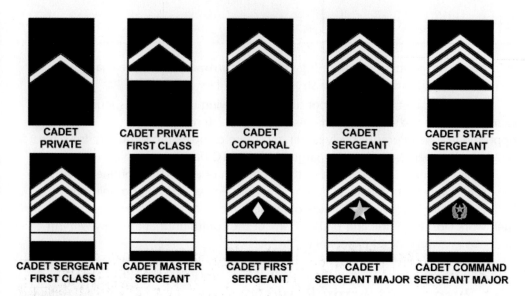

Figure 1.2 Cadet NCO Ranks.

chevrons. Cadet staff sergeants wear three chevrons over a bar. Cadet sergeants first class wear three chevrons over two bars and cadet master sergeants wear three chevrons over three bars.

Cadet first sergeants wear three chevrons over three bars with a diamond between the chevrons and the bars. Cadet sergeants major wear three chevrons over three bars with a star between the chevrons and the bars. Cadet command sergeants major wear three chevrons over three bars with a star circled by a wreath between the chevrons and the bars.

ROTC recognizes three additional cadet enlisted ranks, although these are not non-commissioned officer ranks. Basic cadets wear no insignia. Cadet privates wear one chevron, and cadet privates first class wear a chevron over a bar.

The Cadet Unit Structure

The Army is made of small units—organizational building blocks—which, when combined, create larger units. Within this organizational structure, leaders command and control the resources necessary to deter conflict and to win when combat is necessary.

Your cadet battalion and subordinate-unit structure will vary depending upon the size, makeup, and location of your school, but throughout Army Cadet Command, the unit structures will resemble that in Figure 1.3.

Critical Thinking

What would be the effectiveness of an army without ranks? How well would it operate?

Critical Thinking

What advantages and disadvantages are there to the display of rank insignia in a combat zone?

The Cadet Chain of Command

The cadet chain of command is formed from the MSL IV class at the beginning of each school year. Performance in ROTC, performance at the Leader Development and Assessment Course (LDAC), academics, and overall accomplishments dictate the appointment criteria.

The cadet chain of command does two things:

1. It helps the Professor of Military Science (the PMS) and other cadre accomplish battalion missions and responsibilities.

2. It trains and develops subordinates.

The cadet chain of command includes four unit levels: battalion, company, platoon, and squad.

The explanation below will include the responsibilities of commanders and staff officers, who help the unit accomplish its missions.

At the Battalion Level—

1. The Cadet Battalion Commander:
   a. Commands the corps of cadets, sets the example, and leads the way.
   b. Coordinates with the battalion staff to supervise leadership labs and other events to ensure the training is effective, motivating, and safe.
   c. Conducts meetings and leads battalion formations such as weekly battalion training meetings to coordinate and confirm all training, logistical, and administrative requirements.
   d. Trains and evaluates other cadets—officers, NCOs, and enlisted ranks.

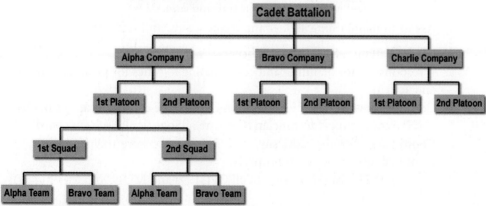

Cadet Bn Organizational Structure Diagram

Figure 1.3 Cadet Unit Structure

2. The Cadet Battalion Executive Officer (XO):
   a. Supervises and coordinates all staff functions.
   b. Commands the battalion in the absence of the battalion commander.
   c. Assists the battalion commander in the performance of his or her duties.
   d. Attends the weekly battalion training meeting.
   e. Provides instruction and evaluation as required.

3. The Cadet Battalion Command Sergeant Major (CSM):
   a. Advises the cadet battalion commander.
   b. Maintains cadet accountability during training.
   c. Checks the cadet NCOs for job knowledge, military appearance, and knowledge of their subordinates' strengths and weaknesses.
   d. Conducts and supervises training to ensure it meets the cadet battalion commander's intent.
   e. Attends battalion training meetings.

4. The Cadet Battalion S1 (administrative officer):
   a. Is responsible for all cadet administration and accountability.
   b. Ensures cadet promotions and absences are documented and managed.
   c. Coordinates, publishes, and executes all cadet social functions and award ceremonies.
   d. Helps the cadre with sponsorship programs.
   e. Provides cadet status reports at each weekly cadet training meeting.

5. The Cadet Battalion S3 (operations officer)—
   a. Is responsible for operations and training in the battalion.
   b. Prepares weekly training meetings and publishes weekly training schedules.
   c. Coordinates all training with the cadre operations officer.
   d. Ensures that all instructors conduct rehearsals and back-briefs.
   e. Provides a training status report at each weekly cadet training meeting.
   f. Coordinates with the cadre operations officer for all required MSL IV cadet evaluations of MSL III cadets.
   g. Publishes all operations orders (OPORDs) and memorandums of instruction (MOIs) on projects, training, and cadet activities.

6. The Assistant Cadet Battalion S3:
   a. Is the primary assistant to the battalion S3.
   b. Assists with instruction for each lab and coordinates with other staff, lab instructors, and cadre to maintain training standards.
   c. Assists in maintaining all training records, aids, and references.

7. The Cadet Battalion S4 (logistics officer):
   a. Is responsible for planning and coordinating logistics for projects, training, and activities.
   b. Coordinates with the cadet operations officer, the cadre operations officer, and the cadre supply technician to ensure all logistics have been coordinated.
   c. Coordinates with the cadre supply technician to ensure 100 percent accountability of battalion property.
   d. Prepares the logistical portion of all OPORDs and MOIs.

8. The Cadet Battalion Public Affairs Officer (PAO):
   a. Coordinates all PAO activities with the cadre S1, university PAO, and community PAO.
   b. Coordinates advertising campaigns, news releases, and feature articles to increase public awareness of ROTC.
   c. Assists in activities involving the ROTC Advisory Council, the ROTC Alumni, the ROTC Hall of Fame, and others.

At Company, Platoon, and Squad Levels—

1. The Cadet Company Commander:
   a. Commands the company and is responsible for its day-to-day operations.
   b. Reports directly to the cadet battalion commander on the morale, welfare, accountability, training, and discipline of the company.
   c. Plans, organizes, and executes company training.
   d. Is responsible for ensuring that the MSL I and II cadets are prepared for their follow-on years of ROTC.

2. The Cadet Company Executive Officer (XO):
   a. Commands the company in the absence of the company commander.
   b. Assists the company commander in the performance of his or her duties.

3. The Cadet Company First Sergeant:
   a. Holds company formations in accordance with Field Manual (FM) 3-21.5 and receives an accurate report from the cadet platoon sergeants.
   b. Supervises cadet accountability during training.
   c. Checks cadet NCOs for job knowledge, military appearance, and knowledge of their subordinates' strengths and weaknesses.
   d. Conducts and supervises training to ensure it meets the cadet company commander's intent.

4. The Cadet Platoon Leader:
   a. Is responsible for the platoon's day-to-day operations.
   b. Is responsible to the cadet company commander on all matters concerning the morale, welfare, accountability, training, and discipline of the platoon.
   c. Plans, organizes, and executes platoon training.

5. The Cadet Platoon Sergeant:
   a. Assists the platoon leader and supervises and coordinates with the squad leaders.
   b. Holds platoon formations in accordance with FM 3-21.5 and maintains accountability for personnel at all times during military functions.
   c. Conducts and supervises training.
   d. Works with the cadet first sergeant on the issue, receipt, and accountability of all equipment and supplies for the platoon.
   e. Acts on the platoon leader's behalf during the platoon leader's absence.

6. The Cadet Squad Leader:
   a. Holds squad formations in accordance with FM 3-21.5.
   b. Checks uniforms and equipment.
   c. Knows each squad member's strengths and weaknesses.
   d. Conducts and supervises training.

Abbreviations for Army Officer Ranks

GEN–general
LTG–lieutenant general
MG–major general
BG–brigadier general
COL–colonel
LTC–lieutenant colonel
MAJ–major
CPT–captain
1LT–first lieutenant
2LT–second lieutenant

CONCLUSION

Effective organization requires both leaders and followers. In the Army, fast and effective decision making requires leadership and teamwork. In combat, there can be no hesitation about who is leading and who is following. Rank takes away the guesswork about who is in charge of what, so leaders and followers can do their jobs.

Learning Assessment

1. What purpose does the division of the Army into officers, noncommissioned officers, and enlisted ranks serve?

2. List the following in order of size: a squad, a company, and a platoon.

3. What is the cadet battalion chain of command?

4. List the cadet ranks.

Key Words

rank
chain of command
unit

References

All Empires: An Online History Community. Retrieved 10 August 2005, from http://www.allempires.com/empires/huns/huns1.htm

AR 670-1, *Wear and Appearance of Army Uniforms and Insignia*. 3 February 2005.

Long, O. F. (1895). *Changes in the Uniform of the Army: 1774-1895*. (Army and Navy Regulation). Washington, DC: Army Quartermaster Corps.

US MILITARY CUSTOMS AND COURTESIES

Key Points

1 Military Customs and Courtesies: Signs of Honor and Respect

2 Courtesies to Colors, Music, and Individuals

3 Military Customs: Rank and Saluting

4 Reporting to a Superior Officer

The courtesy of the salute is encumbent on all military personnel, whether in garrison or in public places, in uniform or civilian clothes. The exchange of salutes in public places impresses the public with our professional sincerity, and stamps officers and enlisted men as members of the Governmental instrumentality which ensures law and order and the preservation of the nation (US Army Military History Institute, 1989).

GEN Hugh Drum

Introduction

A custom is a social convention stemming from tradition and enforced as an unwritten law. A courtesy is a respectful behavior often linked to a custom. A **military courtesy** is such behavior extended to a person or thing that honors them in some way.

Military customs and courtesies define the profession of arms. When you display military customs and courtesies in various situations, you demonstrate to yourself and others your commitment to duty, honor, and country.

As a cadet and future Army leader, you must recognize that military customs and courtesies are your constant means of showing that the standard of conduct for officers and Soldiers is high and disciplined, is based on a code akin to chivalry, and is universal throughout the profession of arms.

military courtesy

the respect and honor shown to military traditions, practices, symbols, and individuals

military customs

those time-honored practices and outwards signs of military courtesy that create a formal atmosphere of respect and honor

Military Customs and Courtesies: Signs of Honor and Respect

Every branch of the armed services has a variety of characteristic customs established long ago and still in use today. Army customs and courtesies lend color, distinction, and ceremony to your daily life as a Soldier.

Courtesies are the outward signs of your respect to your nation, your flag, your comrades, and our country's fallen heroes. They engender mutual respect, good manners, politeness, and discipline.

Customs include such things as responding to a senior officer's presence, recognizing the officer's rank or position of honor, correctly using military titles, wearing headgear, saluting appropriately, reporting correctly, and honoring national and Army symbols and music.

Courtesies to Colors, Music, and Individuals

Courtesies to Colors

National and organizational flags that fly from flagstaffs equipped with finials (the decorative top pieces), are *Colors*. When not in use, colors are furled and encased to protect them. The military detachment assigned to protect, carry, and display them is called a *color guard*. When you pass an unfurled, uncased national Color, salute at six steps' distance, and hold the salute until you have passed six steps beyond the flag. Similarly, if the uncased Color passes by, salute when the flag is six steps away and hold the salute until it has passed six steps beyond.

You shouldn't salute small national flags carried by individuals, such as those carried by civilian spectators at a parade or printed on solid objects. It's also improper for you to salute while you have any object in your right hand or a cigarette, cigar, or pipe in your mouth.

Courtesies to Music

Military music dates back to the early Roman times when such music called military formations together. Now military music establishes a sense of alertness, urgency, attention to detail, self-discipline, and confidence.

Outdoors, whenever and wherever the United States National Anthem, "To the Color," "Reveille," or "Hail to the Chief" is played, at the first note, all Soldiers in uniform and not in formation face the flag—or the music, if the flag is not in view—stand at attention,

The National Colors and the Army Colors are followed by organizational colors.

and give the prescribed salute. Hold the salute position until the last note of the music sounds. Military personnel not in uniform will stand at attention, removing headgear, if any, and place the right hand over the heart. Vehicles in motion come to a halt. Soldiers riding in a passenger car or on a motorcycle dismount and salute. Occupants of other types of military vehicles and buses remain in the vehicle; the individual in charge of each vehicle dismounts and renders the hand salute. Tank and armored car commanders salute from the vehicle.

Indoors, when honoring the US flag, national anthem, and bugle calls, officers and Soldiers stand at attention and face the music or the flag, if one is present.

Other songs worthy of respect and honor include "The Army Song," "Stars and Stripes Forever," "America the Beautiful," and "God Bless America."

Bugle calls are another form of military music to which you should respond with attention. These include:

- Attention, Assembly, Adjutant's Call
- Carry On, Mess Call, Recall
- Taps, Tattoo
- To the Color, National Anthem
- Sound Off, 1st Call
- Reveille and Retreat

Courtesies to Individuals

You show respect for people by standing when they enter a room or enter a conversation. When not at attention or saluting, you allow officers the position of honor at a table. These informal gestures demonstrate your character and respect for Army values.

Uncovering—removing a beret, hat, or headgear—isn't just good manners; it's a sign of respect to others. You should remove your headgear indoors, unless you are under arms. Officers and enlisted Soldiers uncover when they sit as a member of or in attendance on a court or board, when entering places of divine worship, and during attendance at an official reception.

The expression "under arms" means carrying a weapon in your hands, by a sling, or in a holster.

Figure 2.1 Military Rank Progression

Outdoors, you should not remove your military headgear nor raise it as a form of salutation.

Military Customs: Rank and Saluting

Military Rank

For thousands of years, in almost all cultures, leaders have worn or held symbols of their position and authority. In ancient Rome, magistrates carried the *fasces*, a bundle of rods with an axe protruding, as a symbol of their power. Native American chiefs donned eagle feathers to represent their bravery and status in the tribe.

So it's natural for the Army to have rank: Insignia of rank identify who is in charge. The US Army adapted much of its rank structure from the British military tradition. Military rank—unlike a pay grade, which is an administrative feature—is a visible mark of responsibility and leadership meriting recognition and respect. The customary way you recognize an officer of superior rank is to salute him or her.

The Military Salute

The origin of the **military salute** is uncertain, but it probably began as a gesture of trust to show that a person was not holding a weapon. Some historians believe saluting began in Roman times when assassinations were common. A citizen who wanted to see a public official had to approach with his right hand raised to show that he did not hold a weapon. Knights in armor raised their visors with the right hand when meeting other knights. This practice gradually became a way of showing respect and, in early American history, sometimes involved removing one's hat. By 1820, the motion had become touching the hat, and since then the gesture has evolved into the hand salute used in the military today.

the military salute

a formal, one-count military gesture of respect given at attention, in which a subordinate acknowledges a superior officer by bringing the hand to the brim of the cap or to a point slightly above the right eye— Soldiers salute in greeting, leaving, reporting, and other military situations to publicly show respect for the superior's rank

When to Salute

A salute is a public sign of respect and recognition of another's higher rank. When in uniform, you salute when you meet and recognize an officer entitled to a salute by rank except when inappropriate or impractical. Generally, in any case not covered by specific situations, a salute is the respectful, appropriate way to acknowledge a superior officer.

Times you should salute include:

- When the US National Anthem, "To the Color," "Hail to the Chief," or foreign national anthems are played
- When you see uncased National Colors outdoors
- On ceremonial occasions
- At reveille and retreat ceremonies during the raising or lowering of the US flag
- During the sounding of honors
- When the Pledge of Allegiance is being recited outdoors
- When relieving an officer or turning over control of formations
- When rendering reports
- When greeting officers of friendly foreign countries
- When you see officers in official vehicles

When *Not* to Salute

You don't have to salute indoors, except when you report to a superior officer. If either person is wearing civilian clothes and you do not recognize the other person as a superior officer, salutes are unnecessary.

Use common sense. If you are carrying something with both hands or doing something that makes saluting impractical, you are not required to salute a senior officer or return a salute to a subordinate.

Use your judgment. You don't have to salute in an airplane, on a bus, when driving a vehicle, or in public places such as inside theaters or other places of business. The driver of a moving vehicle does not initiate a salute.

Sometimes saluting is inappropriate. Soldiers participating in games and members of work details do not salute. Soldiers reporting to an NCO do not salute.

How to Salute

The hand salute is a smart, one-count movement at the command "Present, arms." When wearing headgear with a visor with or without glasses, on the command of execution "arms," raise the right hand sharply, fingers and thumb extended and joined, palm facing down. Place the tip of the right forefinger on the rim of the visor slightly to the right of the right eye. The outer edge of the hand barely cants downward, so that neither the back of the hand nor the palm is clearly visible from the front. The hand and wrist are straight, the elbow inclined slightly forward, and the upper arm horizontal.

A well-executed salute is crisp, quick, and immediate, with both subordinate and senior officer making the movement in a professional gesture of respect and recognition of that respect. Saluting should become a reflex to you.

Saluting People in Vehicles

You should practice the appropriate courtesy of saluting officers in official vehicles, recognized individually by grade or by identifying vehicle plates and/or flags. You don't salute officers or return salutes from subordinates who are driving or riding in privately owned vehicles.

Greeting an officer

Saluting in Formation

In formation, you don't salute or return salutes except at the command "Present, arms." An individual in formation at ease or at rest comes to attention when addressed by an officer. In this case, the individual in charge salutes and acknowledges salutes on behalf of the entire formation. Commanders of units that are not a part of a larger formation salute officers of higher grade by bringing the unit to attention before saluting. When under battle or simulated battle conditions, you should not call your unit to attention.

Saluting Out of Formation

When an officer approaches a group of individuals not in formation, the first person noticing the officer calls everyone present to attention. All come sharply to attention and salute.

If you are in charge of a work detail, but not actively engaged, you salute and acknowledge salutes for the entire detail.

A unit resting along a road does not automatically come to attention upon the approach of an officer. If the officer speaks to an individual or the group, however, the individual or group comes to attention and remains at attention—unless otherwise ordered—until the conversation ends, at which time the individual or group salutes the officer.

Reporting to a Superior Officer
Reporting Indoors

When **reporting** to a superior officer in his or her office, the cadet, officer, or Soldier removes headgear, knocks, and enters when told to do so, approaches within two steps of the officer's desk, halts, salutes, and reports, "Sir (Ma'am), Cadet Jones reports." Hold the

reporting

the procedure for approaching and speaking to a superior officer that includes approaching, standing at attention, saluting, politely addressing the officer, waiting for recognition, and concisely giving the necessary report, message, or briefing

Critical Thinking

What could be one practical purpose of the reporting procedure? Why might such formal address be necessary to good order, clarity, and precision in the field?

salute until your report is complete and the officer has returned your salute. At the end of the report, you salute again, holding the salute until it is returned. Then you smartly execute the appropriate facing movement and depart. When reporting indoors under arms, the procedure is the same, except that you don't remove your headgear and you render the salute prescribed for the weapon you are carrying.

When a Soldier reports to an NCO, the procedures are the same, except that the two exchange no salutes.

Reporting Outdoors

When reporting outdoors, you move rapidly toward the senior officer, halt approximately three steps from the officer, salute, and concisely make your report, as you do indoors. When dismissed by the officer, you exchange salutes again. If under arms, you should carry your weapon in the manner prescribed for saluting with that weapon. (See FM 3–21.5, Appendix A.)

CONCLUSION

The disciplined exercise of military customs and courtesies in an organization is a clear indicator of the morale and leadership of that organization. It also indicates an organization's ability to function under stress as a team of professionals bound by a warrior ethos and mutual respect.

As an ROTC cadet, you acknowledge your place in the profession of arms through your crisp, professional exercise of customs and courtesies—smartly saluting, properly reporting, and otherwise showing appropriate signs of honor and respect.

Learning Assessment

1. While listening to Army music, songs, and bugle calls, demonstrate your ability to stand at attention, salute, report, and uncover at appropriate times by role-playing with classmates.

2. Name three times when you are required to salute a superior officer and three times when saluting is unnecessary.

3. What is the difference between a flag and a color?

4. What is the difference between a courtesy and a custom? Give an example of each.

5. Explain the purpose of military rank.

Key Words

military courtesy
military customs
the military salute
reporting

References

AR 600–25, *Salutes, Honors and Visits of Courtesy*. 24 September 2004.

Field Manual 670–1, *Wear and Appearance of Uniforms & Insignia*. 3 February 2005.

Field Manual 3–21.5, *Drill and Ceremonies*. 7 July 2003.

US Army Military History Institute. (November 1989). The Hand Salute: A Working Bibliography of MHI Sources. Carlisle, PA. Retrieved 18 July 2005 from http://carlisle-www.army.mil/usamhi/Bibliographies/ReferenceBibliographies/customs/salute.doc

OFFICERSHIP AND THE ARMY PROFESSION

Key Points

1 The Concept of a Profession

2 The Three Characteristics of a Profession

3 Professionalism and the Military

Yours is the profession of arms, the will to win, the sure knowledge that in war there is no substitute for victory, that if you lose, the Nation will be destroyed, that the very obsession of your public service must be duty, honor, country.

GEN Douglas MacArthur

Introduction

The Army requires that you—as a cadet and future officer—accept responsibilities not just for doing a job, but also for assuming a way of life. In other words, the Army requires you to become a professional as stated in FM 1, *The Army*:

> The purpose of any profession is to serve society by effectively delivering a necessary and useful specialized service. To fulfill those societal needs, professions—such as medicine, law, the clergy, and the military—develop and maintain distinct bodies of specialized knowledge and impart expertise through formal, theoretical, and practical education. Each profession establishes a unique subculture that distinguishes practitioners from the society they serve while supporting and enhancing that society. Professions create their own standards of performance and codes of ethics to maintain their effectiveness. To that end, they develop particular vocabularies, establish journals, and sometimes adopt distinct forms of dress. In exchange for holding their membership to high technical and ethical standards, society grants professionals a great deal of autonomy. However, the profession of arms is different from other professions, both as an institution and with respect to its individual members.

As in so much else, GEN George Washington set the example for the Army professional at a time of crisis in our young republic:

GEN Washington at Newburgh

Following its victory at Yorktown in 1781, the Continental Army set up camp at Newburgh, New York, to wait for peace with Great Britain. The central government formed under the Articles of Confederation proved weak and unwilling to supply the Army properly or even pay the Soldiers who had won the war for independence. After months of waiting, many officers, angry and impatient, suggested that the Army march on the seat of government in Philadelphia, Pennsylvania, and force Congress to meet the Army's demands. One colonel even suggested that GEN Washington become King George I.

Upon hearing this, GEN Washington assembled his officers and publicly and emphatically rejected the suggestion. He believed that seizing power by force would have destroyed everything for which the Revolutionary War had been fought. By this action, GEN Washington firmly established an enduring precedent: America's armed forces are subordinate to civilian authority and serve the democratic principles that are now enshrined in the Constitution. GEN Washington's action demonstrated the loyalty to country that the Army must maintain in order to protect the freedom enjoyed by all Americans (FM 22-100).

Critical Thinking

What features of Army officership correspond to the characteristics of a profession, as defined by FM 1?

The Concept of a Profession

When you think about professions, what comes to mind? Credentials, years of study and training, a code of ethics, status in the eyes of the community? Think of the medical caregiver, the legal specialist, the accountant, the architect, the teacher, the law enforcement officer, or the clergy member. What do they have in common with the profession you are training for and are about to enter—the military profession? To truly understand your role as a military professional, you must first understand what people mean when they talk about a *profession*.

A profession is a calling—a vocation. It's a livelihood, yes, but it goes far beyond a simple occupation or "what you do" for a living. A profession is a way of being, a way of thinking, a way of behaving, and a way of growing. In short, it's a *way of life*, not just a job or a lifestyle.

Most professionals would probably tell you that they do what they do because they love their work, that they couldn't do anything else, that they would do it even if they didn't make money at it. The Army is just such a profession. That's why it's important to realize the difference between an occupation—a job—and a profession.

The job of Army officer is the highest embodiment of the profession of arms. When you become a commissioned officer in the US Army, you join an elite body of leaders with a long, proud tradition of service to the country and a commitment to high ideals.

The Three Characteristics of a Profession

One of the outward distinctions of a uniformed profession is to display distinctive clothing. In previous sections, you've read about the importance of military rank, insignia, and the uniform.

Another characteristic of a profession is dedication to service. You perform your service in the military within a profession.

Three characteristics distinguish a profession as a special type of work: **expertise**, **responsibility**, and **corporate culture**.

Critical Thinking

What's the difference between a profession and an occupation?

Expertise

Professionals are who and what they are because they acquire *expertise*—a special kind of knowledge and a context for that knowledge. Expertise can be of at least three types: *technical*—based on training and study; *theoretical/intellectual*—based on education and study; and *liberal*—based on broad reading, interactions with colleagues, and a focus on lifelong learning.

Technical Expertise

Professionals know how to operate the hardware of their professions. They are technically expert. For the doctor, it is an in-depth knowledge of medicines and drugs, laboratory testing, imaging equipment, and surgical procedures. For the architect, it is a well-defined knowledge of loads, stresses, materials tolerances, and coefficients. For the military professional, it is an intimate knowledge of the technology of field weapons, aircraft, tank, artillery, computers, telecommunications, or other specialized equipment.

Theoretical (or Intellectual) Expertise

This is the "how" and "why" of the technical component. In our example of medicine, technicians can perform the tests, but may not know the how and why behind the intricate functions of the human body. A construction worker knows how to pour cement for a reinforced foundation, but may not understand how or why the foundation will hold up the building. The same is true of the Soldier who knows the basic functions of the Army's equipment. It's not necessary for him or her to understand the theoretical concepts behind the tactical or strategic use of the equipment. This component of the professional officer's job is what enables him or her to comprehend and apply new techniques.

Life or death decisions come from an officer's ability to understand the greater mission and apply tactical skills to fulfilling the mission. The only information an officer may receive is the commander's intent. He or she must move forward and deduce the how and why, especially during combat and in today's Contemporary Operating Environment.

Broad Liberal Expertise

This is probably the most complex and the most important component of expertise. Liberal knowledge is a professional officer's ability to understand the role of his or her profession and its unique expertise within society. It includes the knowledge of behavior, human relationships, standards of conduct, and the structures of human organizations. A professional needs to know when and how to offer his services to achieve the most desirable and effective results. You could call this the *philosophy* of arms or the doctrinal grounding of military science.

Military Expertise

What expertise does the Army expect you, as an officer in training, to master? The Army expects you to learn how to organize, equip, and train the force—your Soldiers. The Army also expects you to plan the activities of the force with clarity of mission. You also must know how to execute the mission, a task critical to the military's success. Finally, the Army expects you to be expert in directing operations—engaging in the many kinds of Army activities during both war and peace. Directing operations is a core responsibility of a professional military officer.

---

expertise

what you know (theoretical), what you know how to do (technical), and what the value of that knowledge is to the greater society (broad liberal application)

Responsibility

responsibility

the obligation to employ available information, resources, and personnel to manage work and complete missions, which for an Army professional ultimately means upholding the Constitution and ensuring the security of the nation

Professional Responsibility

As you've seen, professionals require intensive education in a particular service or skill that most members of society do not have. Along with all of this expertise are some distinct responsibilities. As an Army professional, you must be aware of them.

By definition, a professional offers a service that is vital to society. This service is performed for a person or group of people commonly referred to as clients.

By definition, professional expertise is so complex that laymen are usually not capable of understanding what or how the professional does. Therefore, the professional has exclusive possession of a certain skill set, and the client agrees to accept the professional's application of those skills. This relationship creates certain expectations. Just as the professional expects the client to place affairs completely within his or her hands, the client expects the professional to observe certain ethical standards of behavior. Society expects the Army professional to fulfill three key obligations:

- Not exceed professional competence (no "Custer's Last Stands")
- Act only in the nation's best interest
- Maintain integrity with the American people.

Clearly the professional has most of the leverage in this relationship, at least until proven otherwise, and he or she is accepted as the unquestioned authority.

Two major motivating factors prevent abuse of this power. The *first* is the vocation or "calling" aspect of the profession discussed earlier. Most people enter a profession because of an abiding desire to serve society and their fellow man. Many endure great personal hardship to meet the standards of their "calling." Consider the medical student who graduates with high-dollar student loans after spending many years as a full-time student—or your own sacrifices to attend college and become a military professional.

The *second* motivator is autonomy. Most professionals desire to maintain their ability to control their profession. They realize that only as long as the profession as a whole abides by ethical standards will society allow it to keep functioning autonomously.

Responsibility of Officership

The military officer is responsible for the military security of the United States.
Most Americans accept the idea of allowing Army professionals to safeguard and carry out the business of protecting the nation. Most would not know how to fight a war and, when placed in a situation of grave danger, would quickly defer to the military officer for that expertise.

Today, the Army spends billions of dollars and huge amounts of time building competence in the military profession. The structure of the Army rests on ensuring that military professionals do not assume command positions until they are capable of doing so.
The US Constitution, the Soldier's Creed, and the Army Leadership Framework are all sources of the ethical authority and moral obligation to which military professionals adhere in maintaining the trust of the American people and our allies. You will study the Soldier's Creed, Army values, and Army Leadership Framework in other sections of this textbook.

Corporate Culture

In the Army, the corporate culture for military professionals is found, in part, in its customs and courtesies you have already studied, and in the "Warrior Ethos" you will also learn more about in other sections.

What does corporate culture mean? It refers to a group of people experiencing a sense of belonging or a common bond. Among Army professionals, corporate culture tends to result from the following factors:

corporate culture

an organization's internal, unique way of doing things—the fingerprint of the organization, differing from every other organization's culture, binding the organization together from the inside.

- A common bond of mission, shared customs and courtesies, and comradeship
- A desire to remain autonomous
- A unique professional knowledge and expertise.

These factors tend to lead to standard professional practices. Among them are:

1. *A desire to police the profession.* The military has specific doctrine to follow; it also has codes and its own military justice system.

2. *Control of entry.* You must pass the numerous tests used to place military professionals into positions, and you must also go through extensive training and education to receive your commission. A commission is to the officer what a license or certification is to a doctor, attorney, or accountant.

3. *The need to promote professional knowledge.* The key focus is to develop leaders who can fulfill the mission. This requires continuous, extensive training and education.

4. *A desire to represent the profession.* Every action taken by a military professional is representative of the military, down to the uniform he or she wears or the language he or she uses.

Professionalism and the Military

You exhibit professionalism as an Army officer and leader when you respect the Constitution and the military's civilian leadership; when you live the Soldier's Creed and Army values; and when you can apply the elements of the Army Leadership Framework—Be, Know, Do—in all of your daily activities, no matter what your rank, your current job, or where you are assigned.

While most professionals serve an individual client, the Army officer's client is the nation—whether helping in disaster recovery, protecting national security, or defeating an enemy. As GEN MacArthur notes in the quotation at the beginning of this section, the military's professional failure would be catastrophic. Army officers—like their counterparts in the other armed services—study, work, and train throughout their professional careers to ensure that the military profession will not fail when duty calls.

CONCLUSION

The Army officer is the cornerstone of the nation's military. As George Washington, Douglas MacArthur, and other leaders have asserted, without a strong officer corps, the Army lacks the basis of professionalism critical for national security. Being an Army officer is not just another job. It's a proud profession with a rich history of serving the nation. Army training and leader development are unrivaled in the world. Your ROTC program aims to help you become the kind of Army officer—to reach a degree of professionalism—of which you, your family, and our country can be proud.

Learning Assessment

1. Define the concept of a profession.
2. List and define the three characteristics of a profession.
3. Discuss how a doctor, lawyer, and military officer each approach their vocations with the same mindset.

Key Words

expertise
responsibility
corporate culture

References

Field Manual 1, *The Army*. 14 June 2005.

Field Manual 22-100, *Army Leadership: Be, Know, Do*. 31 August 1999.

ORIENTEERING

Key Points

1 Understanding Orienteering

2 Using a Map

3 Finding Your Way

4 Orienteering Terms and Techniques

Today, the complexities of tactical operations and deployment of troops are such that it is essential for all Soldiers to be able to read and interpret their maps in order to move quickly and effectively on the battlefield.

FM 3–25.26

Introduction

As an officer and a Soldier, one of your most important pieces of equipment will be a map.

Knowing how to read that map, knowing where you are, and knowing where you are going allows you to call for indirect fire (for example, artillery support), close air support (such as Army aviation assets), and medical evacuation. Using that map is critical to your survival, your Soldiers' survival, and the success of your mission.

This section has three goals:

- to introduce you to some basic concepts and techniques of **orienteering**
- to introduce you to some basic map reading and land-navigation skills that will help you find your way in unfamiliar territory, such as your college campus or around your ROTC training area.
- to give you a foundation for success as you further develop your map reading and land-navigation skills throughout the ROTC program.

In the following vignette, LTC Robert Ballard studied the terrain features of the French countryside while in flight prior to parachuting into Normandy. His ability to terrain associate and navigate became critical to his survival. As soon as he parachuted to the ground, he compared what he had seen during his flight and descent with his map, determined his location, and continued his mission.

On the Ground in Normandy

It had been the practice of Second Battalion [501st Parachute Infantry Regiment, 101 Airborne Division] to use a large bell and a green electric lantern for assembly following the drop. Coming into Normandy [in France, the night before D-Day, 6 June 1944], these two markers were jumped with personnel. But both of the men were lost and so the assembly ground went unmarked.

LTC Robert A. Ballard came to earth right on the drop zone, which put him about 600 yards to the southeast of Les Droueries. His experience was unique among the battalion commanders of 101st Division in that he knew from the beginning that he was in the right spot. He wasn't quite sure why he knew except that the ground looked as he had expected to find it. Too, he had carefully noted the river courses and roads on the flight in, and when he had jumped, he had felt certain that the calculation had been about right.

Now, lying on the ground, he thought back over the drop and he figured he had probably drifted a little bit. But it was still only a question of being a few fields distant from the point he had been seeking. Mortar and machine gun fire was enlivening the neighborhood; the closest shells were dropping 50 to 75 yards away.

Ballard [had] landed within 25 yards of a hedgerow but he didn't crawl to it immediately. He lay perfectly still for about three or four minutes except for getting a grenade ready while thinking out his next move. He had seen tracer fire follow him during the descent and he strained to know whether he had been spotted. He freed himself at last and ran to a ditch.

There he took out a map and a flashlight and from his reading he knew his location for certain within a few hundred yards; the map checked with what he had remembered of the land picture as he came to earth (History Section, US Army European Theater of Operations).

Understanding Orienteering

FM 3–25.26, *Map Reading and Land Navigation*, describes **orienteering** as:

> . . . a competitive form of land navigation suitable for all ages and degrees of fitness and skill. It provides the suspense and excitement of a treasure hunt. The object of orienteering is to locate control points by using a map and compass to navigate through the woods. The courses may be as long as 10 kilometers (FM 3–25.26).

The American Heritage Dictionary defines orienteering as "a cross-country race in which competitors use a map and compass to find their way through unfamiliar territory."

Orienteering began in Scandinavia in the 19th century as a military event and a part of military training. It became a competitive sport in the early 20th century in Sweden and came to the United States after World War II.

The object of an orienteering competition is to find a series of specific locations (often called *control points* or *targets*) on the ground. Each participant gets a topographic map with the control points circled. The terrain is usually wooded and uninhabited, and allows for different levels of competition. The course setter (the person who plans the course) tries to keep the course interesting, but not so complicated that the competitors can't complete it.

There are several types of orienteering events. The most common are:

- *Route orienteering.* A master competitor leads a group as it walks a route. Beginners trace the route on the map as they walk it on the ground and circle the control points. In another variation, a route is laid out with markers for individual competitors to follow. The winner is the competitor who has successfully traced the route and accurately plotted the most control points.
- *Line orienteering.* Competitors trace their route from a master map and then walk it, circling the control points as they locate them on the ground. The course usually contains five or more control points.
- *Cross-country orienteering.* This most common type of orienteering event is also called *free* or *point* orienteering. Competitors start at one-minute intervals and visit the control points in the same order. The contestant with the fastest time around the course wins. The course usually contains six to 12 control markers.
- *Score orienteering.* Control points are scattered around the competition area. Those near the start and finish point have a low point value, while those farther away have a higher value. Competitors locate as many control points as they can within a specified time, often 90 minutes. Competitors earn points for hitting the control points and lose points for exceeding the specified time. The contestant with the most points wins.

Like any competitive event, an orienteering competition has officials, scorecards, and a start and finish area. Control points are indicated with markers and have some kind of device so that contestants can prove they have visited the control point. The device may be different-colored crayons, punch pliers, letter or number combinations, or stamps or coupons.

To help develop your map reading and navigation skills, your ROTC instructor may set up an orienteering course using a combination of orienteering course types and rules in order to cater your orienteering lesson to your campus and to your freshman experience. You and your fellow cadets, working in teams, should attempt to locate as many control points on your campus map as possible in the time allotted. Below are some of the skills you'll need to successfully complete the course and to prepare for instruction on map reading and land navigation later in your ROTC studies.

Using a Map

The basic tool of orienteering, of course, is a **map**.

Some of the oldest maps still survive on clay tablets archeologists have unearthed in digs of ancient Babylonian cities—present-day Iraq—and date from 2500 to 2300 B.C.E. Demands for better maps came from military necessity. The first tribes needed to map the lay of the land around their villages so they could defend them from other tribes.

Today, maps are everywhere. But for a map to be useful, you must know how to use it.

Orient the Map

Your first step is to orient your map to the north. Almost all modern maps, including most tourist maps, display a north secant arrow somewhere on the map. If a north arrow is not used, as a general rule most maps will show north as "up" or at the top of the map. East is right. West is left. South is "down" or at the bottom of the map. If a north arrow is not used, and the map does not indicate which side of the map is north, then you must orient your map to the lay of the land; that is, you must turn your map so that key buildings, intersections, or terrain features align in the same direction that you are holding your map. This technique is called terrain association, and you will learn more about terrain association in this section as well as in future land navigation lessons.

1. Unfold your map, preferably on a solid flat surface. Familiarize yourself with the map: its size, scale, features, and colors. Read the legend. Locate the north arrow on your map.

2. Open your **compass** and lay it on top of the map. Let the dial of the compass swing freely. (Many compasses lock as you close them. This protects the moving parts.) The magnetic arrow will point north. Rotate your map under your compass until the map's north arrow points in the same direction as the compass arrow. If you do not have a north arrow on your map, and the map legend indicates that north is to the top of the map, then rotate your map until the side

> **map**
>
> *graphic representation of a portion of the earth's surface drawn to scale, as seen from above—it uses colors, symbols, and labels to represent features found on the ground*

> **compass**
>
> *a navigation tool that uses the earth's magnetic field to determine direction*

Critical Thinking

How does learning about a fun and competitive cross-country sport help prepare you to become a better officer?

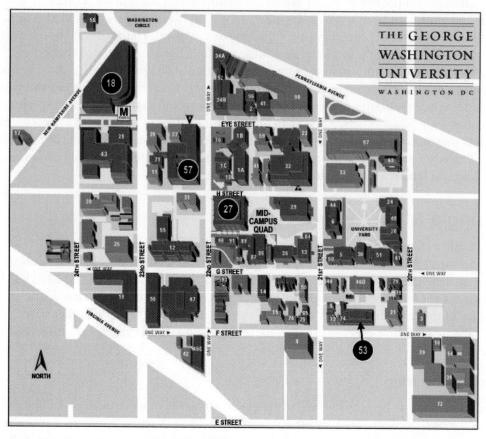

1. **Academic Center**, 801 22nd St.
   A. Phillips Hall
   B. Rome Hall
   C. Smith Hall of Art
   D. Visitor Center
2. **John Quincy Adams House**, 2129-33 Eye Street
3. **Alumni House**, 1925 F St.
4. **Horlense Amsterdam House**, 2110 G St.
5. **Bell Hall**, 2020 G St.
6. **Corcoran Hall**, 725 21st St.
7. **Crawford Hall**, 2119 H St.
8. **Dakota**, 2100 F St.
9. **Davis-Hodgkins House**, 609 21st St.
10. **Abba Eban House**, 607 22nd St.
11. **Fulbright Hall**, 2223 H St.
12. **Funger Hall**, 2201 G St.
13. **Government, Hall of**, 710 21st St.
14. **GSEHD**, 2134 G St.
15. **Guthridge Hall**, 2115 F St.
16. **The George Washington University Club**, 1918 F St.
17. **The George Washington University Inn**, 824 New Hamphsire Ave.
18. **Hospital, GW**, 900 23rd St.
19. **Ivory Towers Residence Hall**, 616 23rd St.
20. **Kennedy Onassis Hall**, 2222 Eye St.
21. **Key Hall**, 600 20th St.
22. **Lafayette Hall**, 2100 Eye St.
23. **Lenthall Houses**, 606-610 21st St.

24. **Lerner Hall**, 2000 H St.
25. **Lerner Framily Health and Wellness Center**, 2301 G St.

Libraries
26. **Jacob Burns (Law)**, 716 20th St.
27. **Melvin Gelman (University)**, 2130 H St.
28. **Paul Himmelfarb Health Sciences (Medical)** 2300 Eye St.
29. **Lisner Auditorium**, 730 21sst St.
30. **Lisner Hall**, 2023 G St.
31. **Madison Hall**, 735 22nd St.
32. **Marvin Center**, 800 21st St.
33. **Media & Public Affairs**, 805 21st St.
34. **Medical Faculty Associates**, 2150 Pennsylvania Ave.
   A. H.B. Burns Memorial Bldg.
   B. Ambulatory Care Center
35. **Mitchell Hall**, 514 19th St.
36. **Monroe Hall**, 2115 G St.
37. **Munson Hall**, 2212 Eye St.
38. **New Hall**, 2350 H St.
39. **Old Main**, 1922 F St.
40. **Quigley's**, 619 21st St.
41. **Rice Hall**, 2121 Eye St.
42. **Riverside Towers Hall**, 2201 Virginia Ave.
43. **Ross Hall**, 2300 Eye St.
44. **Samson Hall**, 2036 H St., 729 21st St.
45. **Schenley Hall**, 2121 H St.
46. **Scholars Village Townhouses**
   A. 619 22nd St.

   B. 2208 F St.
   C. 520-526 22nd St.
   D. 2028 G St.
   E. 605-607 21st St.
47. **Smith Center**, 600 22nd St.
48. **Staughton Hall**, 707 22nd St.
49. **Stockton Hall**, 720 20th St.
50. **Strong Hall**, 620 21st St.
51. **Stuart Hall**, 2013 G St.
52. **Student Health Services**, 2150 Pennsylvania Ave.
53. **Support Building**, 2025 F St.
54. **Thurston Hall**, 1900 F St.
55. **Tompkins Hall of Engineering**, 725 23rd St.
56. **Townhouse Row**, 607 23rd St.
57. **University Garage**, 2211 H St.
58. **Warwick Bldg.**, 2300 K St.
59. **The West End**, 2124 Eye St.
60. **Woodhull House**, 2033 G St.
61. **700 20th St.**
62. **812 20th St.**
63. **814 20th St.**
64. **714 21st St.**
65. **600 21st St.**
66. **609 22nd St.**
67. **613 22nd St.**
68. **615 22nd St.**
69. **617 22nd St.**
70. **837 22nd St.**
71. **817 23rd St.**
72. **9957 E St.**
73. **2033-37 F St.**
74. **2031 F St.**
75. **2101 F St.**
76. **2109 F St.**
77. **2147 F St.**

78. **2000 G St.**
79. **2002 G St.**
80. **2008 G St.**
81. **2030 G St.**
82. **2106 G St.**
83. **2108 G St.**
84. **2112 G St.**
85. **2114 G St.**
86. **2125 G St.**
87. **2127 G St.**
88. **2129 G St.**
89. **2129 G St.** (rear)
90. **2131 G St.**
91. **2131 G St.** (rear)
92. **2136 G St.**
93. **2138 G St.**
94. **2140 G St.**
95. **2142 G St.**
96. **2129-2133 Eye St.** (rear)
97. **2000 Pennsylvania Ave.**
98. **2100 Pennsylvania Ave.**
99. **2136 Pennsylvania Ave.**
100. **2140 Pennsylvania Ave.**
101. **2142 Pennsylvania Ave.**

All addresses are in Northwest Washington.

For assistance or information call the GW Information Center (202) 994-GWGW.

For information on accessibility, call (202) 994-8250 (TDDevice).

Parking
Marvin Center (See #32)
University Garage (See #57)

P Visitor parking entrance

Figure 1.1 Map of The George Washington University
© George Washington University

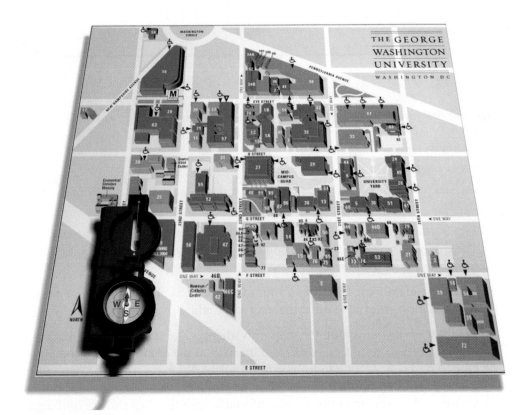

Figure 1.2 Orienting the GWU Map

of your compass is parallel with the side of your map. You have now *oriented* your map to magnetic north. If you have no way of knowing which way on your map is north, then orienting your map to your compass is useless and you must orient your map to the terrain (terrain associate).

Look, for example, at the map of The George Washington University (GWU) in Washington, DC (Figure 1.1). In the lower left-hand corner, you'll see an arrow pointing to north. Using your compass as outlined above, you orient the map so that this north arrow on the map and the compass point in the same direction (see Figure 1.2).

Find Your Location

Now that you have oriented your map, the next step is to figure out where on the map you are located. Doing so is very similar to how you may have determined your location using the maps provided in an amusement park or a shopping mall.

Critical Thinking

Your compass points north because Earth's magnetic field attracts it. Sometimes, however, a compass might not point to true north, or might point a few degrees away from true north. Why would that happen? (You will learn more about the three norths in Section 5.)

Critical Thinking

There are different primitive or "field expedient" methods that you can research on your own as to how to determine your direction both day and night if you do not have a compass. These primitive methods will work, but why is using a compass better?

1. Face north. Look around. Find some identifying features of the area around you. These may be streets and intersections, large buildings (the campus library or the town hall), or hills and streams.

2. Find these identifying features on your map. Spend as much time as you need with this. Be comfortable using the map and finding your location.

3. Compare what you read on the map with what you see on the ground. Locate other features in your area, work from the map to the ground and from the ground to the map. Look in front of you (north). What do you see? What's on the map? Look to your left (west). What do you see? What's on the map? Continue this exercise until you're comfortable with the map and are confident that you have pinpointed your location.

On the GWU map, assume that you have just come up out of the Washington Metro (subway) at the station-marked with an "M" in a square—next to Building 18 in the upper left corner of the map. Standing on 23rd Street facing east, you look to your left (north) and see that the building you are looking at is the GWU Hospital. Looking at the legend underneath the map, you see that Building 18 and the GWU Hospital are the same building. Now you know that you are in the northwest corner of the campus. You further confirm your location by correctly identifying that Building 28 is to your south and Building 20 is to your east. You have pinpointed your location as being at the intersection of 23rd and I (Eye) Street.

Finding Your Way

Once you know where north is and where you are, you're ready to find where you want to go and determine how to get there.

1. *Identify the control points.* (Control points are your targets or the locations you want to go to). These may be already printed on your map or on a plastic (or thin paper) overlay. If the control points are on an overlay, transfer the control points to your map. Mark and identify them with a thin-lead pencil or a thin-point pen. (Keeping the map and the overlay together as you move across campus or through town will be cumbersome.) Ensure that you do not mark over important or identifying features on your map. Your control points may also be named or listed on a separate sheet of paper. If so, you must use your map legend or find your control points on the map using building names, street names, etc. Once you have found your named control points, you can mark them on your map with a light circle or X.

Looking at the GWU map, assume your first control point is at the Support Building. That is your control point. Looking at the alphabetical listing of buildings on campus, you find that it is at 2025 F Street, Building 53. Looking at the map, you find that Building 53 is in the southeast corner of campus. You also notice that the north-south cross streets are numbered (20th, 21st, 22nd, etc.) with the numbers increasing as you move west. The east-west cross streets are lettered (F, G, H, etc.) with the letters deeper in the alphabet as you move north. The diagonal streets are named after states (Pennsylvania, Virginia, New Hampshire).

2. *Plan your route.* Most orienteering events will be timed; the individual or team that returns within the prescribed time with the most correct control points will be determined the winner(s). Because of this, it is crucial for you to correctly plot (mark) your control points accurately on your map and determine the most time-efficient route to take to find as many control points, while still allowing time to return within the time limit. In some orienteering competitions, the sequence in which you must locate your control points may be dictated for you. Plan the course you intend to take to get to each of the control points. You can write down the sequence you wish to take, or you can write your route on your map. One technique to prevent confusing the many routes to control points is to number each "leg" of the route.

   On the GWU map, you have a number of options. You could take I (Eye) Street east two blocks to 21st Street, then turn right (south) and go three blocks to F Street, turn left (east again) and walk to the Support Building. Or, you could walk three blocks south down 23rd Street to F Street, turn left, and walk two and a half blocks east. (Note that the blocks are not all the same size.)

3. *Consider time, distance, crowds, and traffic.* Weigh straight-line routes against ease of passage. (Cutting across the football field may be a good idea as long as the band isn't practicing on the field. You may need to go around.) Also keep safety in mind. Crossing a four-lane highway may be the most direct way to reach a control point, but the risk far outweighs the time you will save.

   Looking at the GWU map, you see the Mid-Campus Quad on H Street between 21st and 22nd Streets. You see that you could cut over from H Street through the Quad to 21st Street and save yourself a few steps. Note also that some of the streets through the campus are one-way streets. You'll want to remember that if you have to drive around campus.

4. *Calculate rough distances.* If your map has a graphic bar scale (similar to a ruler), or some other method to measure distance, you can compute the distance on the map to the distance on the ground. This can be useful in deciding which route to take if a there are many possible routes to a control point. If your map has no method to scale distance (the map is not to scale), then you will need to terrain associate to get a good feel to the scale of your map compared to the actual terrain.

   The GWU map doesn't have a scale, but after you've walked one block, you'll have a feel for the distance on the ground compared to the distance on the map.

5. *Follow your route.* The fastest way to navigate on an orienteering course is to terrain associate. As you move across the ground, compare your map with key features on the ground and keep your pace with the distance on your map. Compare the buildings and terrain around you with the markings on your map.

   As you walk across campus, you turn or rotate your map to keep it oriented toward your direction of travel. If you are traveling north, then your map should be oriented north. If you take a right turn, then you should rotate your map to the right so your map is now oriented east—the direction of your travel. This way, the buildings you see in front of you and alongside will be the same as those shown on the map. The time to use your compass is when you have forgotten to maintain your map's orientation to the lay of the land, or if you are uncertain of your location, or if you simply want to double-check yourself. Pull out your compass and orient your map to your compass just as you did when you first began. Once your map is oriented, you can pick up where you left off. If your map does not have a north arrow, and you are uncertain of your location, you must back track to your last confirmed or known point on the map, or the most prominent feature that you can identify (such as an intersection, bridge, or a major building). Once you have your bearings, continue along your planned route to your next control point.

6. *Locate the control point.* For your ROTC orienteering exercise, your control points may be important campus buildings or facilities that may be easily identifiable as you approach the control point. However, in orienteering competitions, control points aren't always the obvious landmarks that are easily identifiable from a distance. If this is the case with one of your control points, then you must navigate to a landmark or feature—such as a hilltop—as a checkpoint (sometimes called an attack point) to find your nearby control point. Find the control point on your map. Find a nearby checkpoint on your map. Move quickly to the checkpoint, which will be more easily identifiable than the control point. Then find and move to the control point. Checkpoints may be major buildings, stream junctions, bridges, or road intersections.

The Quad itself is a good checkpoint. You can also compare the street signs at each corner with the names of the streets on the map. As you walk east along H Street, check to see that you pass the University Garage (Building 57) on your left. Then, just after you cross 22nd Street, look for the Melvin Gelman University Library (Building 27) on your right. These serve as additional checkpoints. The Quad is just east of the library.

7. *Complete the course and return to the start or rally point.* As described earlier, if your orienteering event is a timed event, you may find yourself in a time crunch and unable to find some of your final few control points. If this is the case, you must adjust your final route(s) in a manner that allows you to find as many of your remaining control points but still arrive back at the finish point within the time limit. In order to adjust your final route(s), you may begin at your last control point, or at an easily identifiable checkpoint. From this location, you must consider distance and time remaining to determine which control points you can find and still return before time runs out. You plan your route (or legs) to those remaining control points that you can find within the prescribed time.

You arrive at the Support Building, check the address and the sign outside the building to make sure you've arrived at the right place.

Congratulations—you're an orienteer!

Orienteering Terms and Techniques

To improve your orienteering skills, learn and use these terms and techniques:

Dead reckoning is moving a set distance along a set line. Generally, it involves moving so many yards or meters in a specific direction, usually a compass reading in degrees.

("Move 350 meters due north" or "Move 1,500 meters on an azimuth [or reading] of 220 degrees.")

As you move along your set line, you may want to identify **steering marks** to guide you. Find a point in the distance—a building, a hilltop, a large tree—on your line and move toward that point.

Dead reckoning has two advantages: It is easy to teach and learn, and it is an accurate way of moving from one point to another over short distances.

Handrails. Find existing linear features—trails, fences, roads, streams, power lines—that parallel your route. Use these "handrails" as a check between control points. On the GWU map, the streets are your handrails.

Pacing. You need to know how to measure distance on the ground. Measure out a (or use an established) 100-meter pace course. Walk the course. Count the number of paces it takes for you to walk the 100 meters. This number is your pace count. Some people use a double-pace count: they step off on their left foot and count every time their left foot hits the ground. It is easier to use a double-pace count because you are counting half the number of steps. Everyone's pace count is different, so never rely on a friend's pace count. The longer your legs, the shorter your pace count, and vice versa. Your pace count (double-pace count) may be 73; that is you reach 100 meters on the 73rd alternate footstep. Your friend, who may be just a few inches taller, may have a much longer stride and may be able to cover the same 100 meters in just 67 paces. It is also important for you to know your pace count for the fractions of 100 meters, such as 25 meters and 33 meters. Knowing paces for these distances will allow you to easily figure out fractions of 100 meters such as 25, 33, 50, 66, and 75 meters without a lot of mental computing.

Terrain association is movement by landmarks. You compare what you see on the ground with what you see on your map as you move. When navigating by terrain association, you must constantly orient your map as you change directions. Moving by terrain association is more forgiving than dead reckoning. If you make a mistake by dead reckoning, many times you must move back to your last known position and begin the dead reckoning over again. With terrain association, you can always quickly find your location by comparing what you see around you with what you see on the map. In most cases, you can identify your location without ever having to backtrack at all. Because of this, terrain association is often less time-consuming than dead reckoning. The example in this section used terrain association to travel across campus using the GWU map.

Thumbing. Thumbing is a technique used in terrain association in which you fold your map small enough to put your thumb next to your start point. Do not move your thumb from your start point. To find your new location, look at your map and use your thumb as a reference for your start point. That way, you don't have to keep looking over the entire map.

---

dead reckoning

a navigation technique by which you travel a set distance (usually in meters) along a set line (usually a compass reading in degrees)

steering marks

landmarks located on the azimuth to be followed in dead reckoning—steering marks are commonly on or near the highest points visible along the azimuth line and are selected based on what you see on the actual terrain, not from a map study. They may be uniquely shaped trees, rocks, hilltops, posts, towers, and buildings— anything that can be easily identified

terrain association

a navigation technique by which you move from one point to another using landmarks and terrain features

---

Critical Thinking

What do you think are the advantages and disadvantages of dead reckoning compared to the advantages and disadvantages of terrain association?

CONCLUSION

Orienteering is a fun way to learn the different land navigation methods and techniques available to you. Your ROTC orienteering lesson is an enjoyable way to introduce you to using a map and a compass to navigate from one point to another. Your ROTC orienteering lesson should allow you to become more familiar with offices and organizations on your campus that can help you during your transition as a college student. As you continue with ROTC, you will apply the knowledge from this orienteering lesson to more advanced skills in map reading, navigation, and terrain analysis. If you enjoyed this lesson on orienteering, you may want to consider further research on competitive orienteering that may be available at or near your college or university by visiting http://www.us.orienteering.org/

While orienteering can be an enjoyable pastime, the map-reading and land-navigation skills it teaches are important life skills for Soldiers and the officers who lead them. In the vignette at the beginning of this section, LTC Ballard knew that he had drifted from his original drop zone or point. His use of orienteering skills—linking what was on the ground to what was on the map—made a difference in his ability to carry out his mission in the crucial first hours of the D-Day invasion.

Learning Assessment

1. Explain how to orient a map to the north.
2. Explain how to relate the points on a map to the points on the ground.
3. Describe how to find a control point on campus, in town, or nearby.

Key Words

orienteering
map
compass
dead reckoning
steering marks
terrain association

References

Field Manual 3–25.26, *Map Reading and Land Navigation*. 18 January 2005.

History Section, US Army European Theater of Operations. (n.d.) Regimental Unit Study 2. *The Fight at the Lock*. File No. 8-3.1 BB 2. Retrieved 30 June 2005 from http://www.army.mil/cmh-pg/documents/wwii/lock/lock.htm

INTRODUCTION TO TACTICS I

Key Points

1 The Elements of a Fire Team

2 The Elements of a Rifle Squad

Soldiers with sharply honed skills form the building blocks of combat effective squads and platoons. They must maintain a high state of physical fitness. They must be experts in the use of their primary weapons. They must be proficient in infantry skills (land navigation, camouflage, individual movement techniques, survival techniques, and so forth). Finally, they must know and practice their roles as members of fire teams, squads, and platoons.

FM 7-8

Introduction

The Army's smallest maneuver element controlled by a leader is the **fire team**. The fire team is the building block for all Army tactical operations. Fire teams make up **squads**; squads make up **platoons**. Army lieutenants lead platoons as part of an infantry rifle company.

This section explains the elements, weapons, roles, and responsibilities of the fire team and rifle squad. Success in tactical operations depends on Soldiers at all levels understanding their tactical mission and the steps necessary to accomplish it.

SGT Tommy Rieman led his rifle squad in a firefight in Iraq in 2003. These Soldiers faced enormous odds—and won.

Beating 10:1 Odds, Soldier Earns Silver Star

August 26, 2004—In a fight, two against one is bad odds. Ten against one is a recipe for disaster. Yet those were the odds SGT Tommy Rieman and his squad faced and beat when they were ambushed by more than 50 anti-American insurgents in Iraq last December.

Rieman, 24, a team leader in Company B, 3rd Battalion, 504th Parachute Infantry Regiment, 82nd Airborne Division, was awarded the Purple Heart and the Silver Star for his heroic actions [that] December day during a ceremony at Devil Brigade Field August 6. He was also awarded the Army Commendation Medal with a "V" device for valor for a separate reconnaissance mission that took place in March 2003.

Rieman was in charge during the patrol that garnered him the Silver Star because he had scouted the area before and knew the terrain. His eight-man patrol was in three light-skinned Humvees with no doors when the first rocket-propelled grenade hit.

"The thing I remember most was the sound of the explosion. It was so loud," said Rieman.

They were hit by three RPGs [rocket-propelled grenades] and a barrage of small arms fire coming from 10 dug-in enemy fighting positions. Staying in the kill zone meant certain death, so the vehicles never stopped moving. Rieman knew he had to return fire. Bullets whizzed after them as the vehicles sped away from the ambush and the Soldiers found themselves caught in another ambush.

There were maybe 50 enemy attackers blasting away at him with small arms fire from a grove of palm trees nearby. Injuries to his men were beginning to pile up. Out of [Rieman's] squad, SGT Bruce Robinson had lost his right leg in the RPG attack and SPC Robert Macallister had been shot in the buttocks. Rieman himself had been shot in the right arm and chest, and had shrapnel wounds to his chest, stomach, and ear. Worst of all, they were almost out of ammo.

He began firing away with his M203 grenade launcher, raining round after round down on the attackers. After being battered by 15 of Rieman's 40mm grenades, the enemy's guns were silent (US Army, 2004).

The Elements of a Fire Team

Sergeant Rieman led an infantry rifle squad of eight other Soldiers. That nine-Soldier squad fought as two four-Soldier fire teams, plus the squad leader, SGT Rieman.

A squad's fire teams are referred to as the Alpha Team and Bravo Team. Each fire team has four Soldiers—a fire team leader, a rifleman, an automatic rifleman, and a grenadier. The team members' rank and experience will range from a private (E-1) straight out of Initial Entry Training (IET) and Advanced Individual Training (AIT) to a specialist (E-4) who may have anywhere from one to four years of experience. The team leader is generally a sergeant (E-5) with three to five years of experience.

The team members' positions (rifleman, automatic rifleman, and grenadier) are based on their assigned weapons (rifle, squad automatic weapon [SAW], and M203 grenade launcher). The platoon leader and platoon sergeant consider each Soldier's experience, skill with the weapon, and longevity in the unit before assigning the Soldier a position on the fire team. The rifleman, team leader, and squad leader carry rifles from the M16/M4 family of weapons, based on the table of organization and equipment (TO&E) for various units.

Rifleman

The *rifleman* carries an M16/M4 rifle, a night-vision device, and an infrared (IR) aiming device. Riflemen carry close-combat optic devices and reflexive fire optics. Their role is to engage targets within the range and capability of their weapon. They may also serve as pace man, compass man, near- or far-side security, en route recorder, or ammo bearer.

The M4 Carbine is the fourth generation of the Army's M16 rifle, which was introduced during the Vietnam War. It fires a 5.56 mm round. The M4 has been a part of the Army inventory since 1997. It uses a 30-round magazine and offers the rifleman two firing modes: semiautomatic (a single shot every time the rifleman pulls the trigger) and three-round burst (three shots every time the rifleman pulls the trigger). It has a shorter barrel than

If the platoon does not have enough sergeants to fill all the platoon's team leader positions, the most experienced specialist with the best leadership qualities will be promoted to the leadership rank of corporal and will lead a fire team. The corporal remains at the pay grade of a specialist (E-4), but has the additional leadership responsibilities of a sergeant. The corporal will remain the fire team leader until promotion to sergeant, or until the platoon receives a new sergeant.

Fire Team

Team Leader

Rifleman Automatic Rifleman Grenadier

Figure 2.1 Fire Team

Figure 2.2 M4/M4A1 Carbine

Figure 2.3 M4 MWS Carbine

Figure 2.4 Mounting the M68 Close Combat Optic to the M16A1/A2/A3

Figure 2.5 M68 Close Combat Optic Mounted on M16A4/M4-Series Weapons

Figure 2.6 Mounting Thermal Weapons Sight on M16A1/A2/A3

the M16 and a collapsible stock, making it lighter and easier for Soldiers to use in close quarters, such as inside a building or in an urban area. Its enhanced upper rail allows Soldiers to mount various day and night sighting devices to improve their effectiveness.

The M4 weighs about seven and a half pounds. A bayonet can be attached for hand-to-hand fighting. It can reach out—accurately—to 600 meters—or more than a quarter-mile away.

Automatic Rifleman

The *automatic rifleman* carries an M249 Squad Automatic Weapon (SAW), a night-vision device, and an IR aiming device. The automatic rifleman's role is to use his or her weapon to maximum effect.

The M249 Squad Automatic Weapon delivers accurate, lethal, and direct automatic fire. Like the M4, it fires a 5.56 mm round. It has been part of the Army inventory since 1987.

TABLE 2.1	Characteristics of the M16-/M4-Series Weapons			
Characteristic	**M16A1**	**M16A2/A3**	**M16A4**	**M4**
WEIGHT (pounds):				
Without magazine and sling	6.35	7.78	9.08	6.49
With sling and loaded:				
20-round magazine	6.75	8.48	9.78	7.19
30-round magazine	7.06	8.79	10.09	7.50
Bayonet knife, M9	1.50	1.50	1.50	1.50
Scabbard	0.30	0.30	0.30	0.30
Sling, M1	0.40	0.40	0.40	0.40
LENGTH (inches):				
Rifle w/bayonet knife	44.25	44.88	44.88	N/A
Overall rifle length	30.00	39.63	39.63	N/A
Butterstock closed	N/A	N/A	N/A	29.75
Butterstock open	N/A	N/A	N/A	33.0
Operational Characteristics:				
Barrel rifling-right hand 1 twist (inches)	12	7	7	7
Muzzle velocity (feet per second)	3,250	3,100	3,100	2,970
Cyclic rate of fire (rounds per minute)	700-800	700-900	800	700-900
Maximum Effective Rate of Fire:				
Semiautomatic (rounds per minute)	45-65	45	45	45
Burst (3-round bursts)(rounds per minute)	N/A	90	90	90
Automatic (rounds per minute)	150-200	150-200 A3	N/A	N/A
Sustained (rounds per minute)	12-15	12-15	12-15	12-15
RANGE (meters):				
Maximum range	2,653	3,600	3,600	3,600
Maximum effective range:				
Point target	460	550	550	500
Area target	N/A	800	600	600

Figure 2.7 M249 Squad Automatic Weapon

CS stands for 0-chloro-benzalmalononitrile, which is a white solid powder usually mixed with a dispersal agent, like methylene chloride, that carries the particles through the air.

Figure 2.8 M4 Carbine with M203A1 Grenade Launcher

The M249 SAW uses 30-round M4 magazines or 200-round pre-loaded plastic magazines. It weighs a little more than 16 pounds, or twice the weight of an M4. The M249 has a maximum effective range of 1,000 meters—a full kilometer or more than half a mile away.

Grenadier

The *grenadier* carries an M203A1 grenade launcher attached to an M4. It adds about 11 pounds to the weight of the M4. The grenadier carries the same equipment as the others in the squad.

The M203A1 grenade launcher fires out 40 mm grenade rounds. It can fire high-explosive (HE) rounds, tear gas (CN/CS/OC) rounds, smoke rounds, non-lethal projectiles, signal rounds, and practice rounds. The maximum effective range for the M203A1 is 350 meters—or three-and-a-half football fields.

Fire Team Leader

The fire team leader carries an M4 and leads by example. The leader moves the fire team and controls the rate and placement of its fire. Fire team leaders keep track of their teams and their teams' equipment. They make sure their teams meet unit standards and help their squad leaders as necessary.

Buddy Teams

Fire teams are divided into two-Soldier *buddy teams*. Buddy team members support and watch out for each other during combat or other operations. FM 7-8 states that the leader and the automatic rifleman form one buddy team and the grenadier and rifleman form the other, but this is not always the case. Based on unit standing operating procedures (SOPs) and depending on the mission, many units may place the grenadier with the team

TABLE 2.2	General Data for M249 Squad Automatic Weapon
Ammunition	5.56-mm ball and tracer (4:1 mix) ammunition is packaged in 200-round drums, each weighing 6.92 pounds; other types of ammunition available are ball, tracer, blank, and dummy.
Tracer burnout	900 meters (+)
Length of M249	40.87 inches
Weight of M249	16.41 pounds
Weight of tripod mount M122 with traversing and elevating mechanism and pintle	16 pounds
Maximum range	3,600 meters
Maximum effective range	1,000 meters with the tripod and T&E
Area: Tripod	1,000 meters
Bipod	800 meters
Point: Tripod	800 meters
Bipod	600 meters
Suppression	1,000 meters
Maximum extent of grazing fire obtainable over uniformly sloping terrain	600 meters
Height of M249 on tripod mount M122A1	16 inches
Rates of Fire:	
Sustained	100 rounds per minute Fired in 6- to 9-round bursts with 4 to 5 seconds between bursts (change barrel every 2 minutes)
Rapid	200 rounds per minute Fired in 6- to 9-round bursts with 2 to 3 seconds between bursts (change barrel every 2 minutes)
Cyclic	650 to 850 rounds per minute Continuous burst (change barrel every minute)
Basic load, ammunition	1,000 rounds (in 200-round drums)
Elevation, tripod controlled	+200 mils
Elevation, tripod free	+445 mils
Depression, tripod controlled	–200 mils
Depression, tripod free	–445 mils
Traverse, controlled by traversing and elevating mechanism	100 mils
Normal sector or file (with tripod)	875 mils

Note: T&E=traverse and elevation mechanism

TABLE 2.3	Technical Data for the M203/M203A1 Grenade Launcher

WEAPON

Length:
Rifle and grenade launcher (overall)	99.0 cm (39 inches)
Barrel only	30.5 cm (12 inches)
Rifling	25.4 cm (10 inches)

Weight:
Launcher, unloaded	1.4 kg (3.0 pounds)
Launcher, loaded	1.6 kg (3.5 pounds)
Rifle and grenade launcher, both fully loaded	5.0 kg (11.0 pounds)

Number of lands:	6 right hand twists

AMMUNITION

Caliber	40 mm
Weight	About 227 grams (8 ounces)

OPERATIONAL CHARACTERISTICS

Action	Single shot

Sights:
Front	Leaf sight assembly
Rear	Quadrant sight

Chamber pressure	206,325 kilopascals (35,000 pounds per square inch)
Muzzle velocity	76 meters per second (250 feet per second)
Maximum range	About 400 meters (1,312 feet)

Maximum effective range:
Fire-team sized area target	350 meters (1,148 feet)
Vehicle or weapon point target	150 meters (492 feet)

Minimum safe firing range (HE):
Training	130 meters (426 feet)
Combat	31 meters (102 feet)

Minimum arming range	About 14 to 38 meters (46 to 125 feet)
Rate of fire	5 to 7 rounds per minute
Minimum combat load	36 high-explosive rounds

leader so that the team leaer can direct the grenadier to mark targets. Because the M203 can mark targets, some units assign M203s to squad leaders so that they can mark targets for the squad.

Why These Weapons?

At this point, you may be wondering why rifle team members carry different weapons. Why not have everyone carry a grenade launcher, for example?

There are several reasons. The fire team must be able to engage and destroy the many different types of targets they are likely to meet on the battlefield. The different weapons the members carry give the team a balance of firepower, simplify the logistics of supplying ammunition, and vary the load each Soldier has to carry. The heavier the weapon, the less other gear or ammunition a Soldier can bear and the less he or she can maneuver.

Critical Thinking

Considering the characteristics, capabilities, and versatility of the fire team's weapons, what factors must the squad leader or fire team leader take into consideration when deciding which team member to assign to each weapon?

The team uses its M16/M4s to engage individual enemy fighters, especially at close quarters, while the M249 is used to engage a concentration of troops or targets covering a larger area with heavy, continuous fire.

Once you compare the maximum rates of fire, the sustained rates of fire, and the maximum range of the rifle and automatic rifle, you can easily understand the value of an automatic rifle on the fire team. The M249's rate of fire and range allow the gunner to cover fire team members when they are maneuvering on the offensive. It is also the backbone of the team's defense when attacked—leaders position the SAW to protect a unit's front, flanks, and rear. In the defense, leaders assign the M249 a *final protective line* to defeat the enemy's attempts to overrun their defensive position. The automatic rifleman has increased lethality (deadliness) and range over the rifleman, but because of the M249's high rate of fire, the automatic rifleman must also carry more ammunition than the rifleman does.

The M203 grenade launcher is the most versatile of the fire team's weapons. The primary advantage the M203 has over the rifle and automatic rifle is that it provides the fire team with indirect fire capabilities; that is, the grenadier can engage targets that he or she cannot see along a direct line of sight. Because the M203 round travels at a high trajectory, or arch, the rounds can reach enemy soldiers in "dead space" (gullies, ditches, trenches, or other fighting positions), or it can reach enemy fighters taking cover in bunkers or in buildings.

When shooting the high-explosive (HE) round, the grenadier is essentially sending out to 350 meters an exploding round with the lethality of a hand grenade—but at a distance far greater than any Soldier can throw a hand grenade. The dual purpose HE round can penetrate up to two inches of light armor, and therefore can destroy trucks and other lightly armored vehicles. During night operations, the grenadier can illuminate the enemy with the star parachute round, allowing other fire team members to see and engage the enemy. The grenadier can show the team members where the enemy is located by using the marking round. The marking round can also be used to indicate landing zones for helicopters or targets for close air support aircraft to attack. The grenadier can also use the star cluster

I love the infantry because they are the underdogs. They are the mud-rain-frost-and-wind boys. They have no comforts, and they even learn to live without the necessities. And in the end they are the guys that wars can't be won without (Tobin, 1987).

Ernie Pyle, World War II correspondent

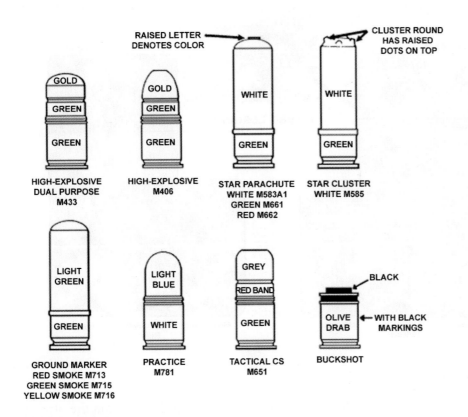

Figure 2.9 Cartridges for the M203 Grenade Launcher

as a prearranged signal to begin a certain tactical action. In urban environments, the grenadier can use tactical CS to flush the enemy out of hiding, or the grenadier can use the buckshot round when entering or clearing rooms or buildings.

As you can see, the various ammunition available for the M203 gives the small fire team a lot of varied firepower and capability that a rifle and an automatic rifle can't provide. The grenadier must also carry a heavier load, however, due to the weight of each round.

The Elements of a Rifle Squad

Two fire teams and a squad leader make up the squad. The rifle squad leader is generally a staff sergeant (E-6) with six to eight years of experience, who came up from the ranks and started as a fire-team rifleman. The squad leader is responsible for everything the squad does—or fails to do. He or she is a tactical leader and leads by example.

Among other things, the squad leader:

- Maneuvers the squad and controls the rate and distribution of its fire
- Trains the squad members in their individual and collective tasks
- Manages the squad's needs, requesting and issuing ammunition, water, rations, and equipment
- Keeps track of the squad's Soldiers and their equipment
- Inspects the squad's weapons, clothing, and equipment and directs their maintenance.

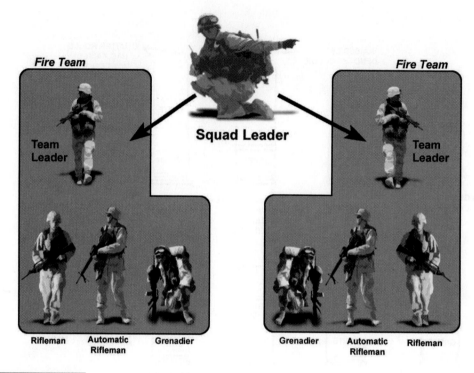

Fire Team

Squad Leader

Team Leader

Team Leader

Rifleman   Automatic Rifleman   Grenadier

Grenadier   Automatic Rifleman   Rifleman

Figure 2.10 Squad

Most of the maneuver and firepower of a rifle platoon derives from the platoon's three rifle squads. Later you will learn more about the infantry rifle platoon and the composition, duties, and responsibilities of the platoon's headquarters element. Together, the rifle squads and headquarters element make up the rifle platoon an Army lieutenant leads. Soon you will study the many types of missions the infantry fire team and squad may be called upon to complete as part of the infantry rifle platoon.

Critical Thinking

How do experience and time in service enhance the rifle squad leader's leadership?

The noncommissioned officer (NCO) corps is called "the backbone of the Army." Why do you think that is?

What can a second lieutenant expect to learn from his or her NCOs?

Critical Thinking

Why do all commissioned officers, regardless of branch, need to know infantry tactics?

CONCLUSION

The infantry is often described as "the tip of the spear." The four Soldiers in a fire team and the nine Soldiers in a rifle squad represent the sharpened tip of that spear. Success in a mission depends on all Soldiers in the fire team and squad understanding the unit's mission—its task and purpose—and how each Soldier's assigned weapon and each Soldier's role and responsibility relate to mission success. Fire teams and squads form the Army's basic maneuver element led by the Army lieutenant—the infantry rifle platoon—the foundation on which all tactical operations rest. Now that you understand the composition and leadership of the fire team and squad, in the next lesson you'll learn how individuals and fire teams move on the battlefield.

Learning Assessment

1. Name the duty positions that make up a fire team.
2. Identify the weapons that each member of the fire team carries.
3. Name four responsibilities of the rifle squad leader.
4. Describe the primary purpose of each weapon in the fire team.

Key Words

fire team
squad
platoon

References:

Field Manual 3-22.9, *Rifle Marksmanship*. Change 3. 28 April 2005.

Field Manual 3-22.31, *40 mm Grenade Launcher, M203*. 13 February 2003.

Field Manual 3-22.68, *Crew Served Machine Guns, 5.56 and 7.62 mm*. 31 January 2003.

Field Manual 7-8, *Infantry Rifle Platoon and Squad*. Change 1. 1 March 2001.

Tobin, J., ed. (1987). *Ernie Pyle's War: America's Eyewitness to World War II*. New York: Simon and Schuster.

US Army. (26 August 2004). Beating 10:1 odds, Soldier earns Silver Star. *Soldier Stories*. Retrieved 15 July 2005 from http://www4.army.mil/ocpa/soldierstories/story.php?story_id_key=6307

INTRODUCTION TO TACTICS II

Key Points

1 **The Three Individual Movement Techniques**

2 **The Two Fire Team Movement Formations**

Infantry platoon and squad leaders must be tacticians. They cannot rely on a book to solve tactical problems. They must understand and use initiative in accomplishing the mission. . . . The art of making sound decisions quickly lies in the knowledge of tactics, the estimate process, and platoon and squad techniques and procedures.

FM 7-8

Introduction

From the introduction of the muzzle-loaded smoothbore musket until the mid-19th century, infantry units in Europe and the Americas **maneuvered** in long lines and large formations. The idea was to subject the enemy, who stood or knelt in similar formations, to massed fire at short range. Often these formations got off one shot, then charged the enemy with fixed bayonets.

That changed during the American Civil War with the advent of the rifled musket, conical bullet, repeating rifle, and primitive machine gun. Infantry tactics did not keep pace with these advances in weaponry, and this led to the increased carnage at battles like Fredericksburg, Gettysburg, and Cold Harbor.

The bloodbaths on World War I battlefields, where British, French, and German infantry charged futilely into machine-gun and grenade fire with massive losses, showed the gap between increasingly lethal military technology and outdated infantry tactics.

Fortunately for the Soldiers of today, much has changed. The modern, trained Army uses traveling techniques, movement formations, and **cover** and **concealment** to advance on or defend objectives with the fewest possible casualties. In this section, you will learn how to move under fire, as an individual and as part of a fire team.

In his book, *Steel My Soldiers' Hearts*, COL David Hackworth relates how his battalion surgeon, CPT Byron Holley, welcomed a young medic to Vietnam:

maneuver

employment of forces in the battlespace through movement in combination with fires to achieve a position of advantage in respect to the enemy in order to accomplish the mission

cover

protection from the effects of direct and indirect fires

concealment

protection from observation and surveillance

Learning to Crawl

Holley couldn't help remembering his own baptism by fire and told Billy how when he heard the bullet snapping by barely a foot above his head, "it was the first realization I had that, hey, a guy can get killed pretty easy over here. I looked up at the moon and prayed, 'God, please don't let me die in this hellhole.' And it was just like I heard a voice saying: Relax, everything's going to be just fine, just remember what you learned in basic training—when the lead's flying, get your butt down. It was like a protective shield came around me and I lost any fear. And I learned fast that you can cover a lot of territory crawling (Hackworth, 2002).

The Three Individual Movement Techniques

As Holley learned, knowing how to move on the battlefield is the key to staying alive. But before you move, you must know where you want to move to next. Stay on the route that your leader selected for the team. Then identify the next covered or concealed position that is nearby. Select your route to your next position so you are exposed to the least amount of enemy fire. And don't forget—you don't want to cross in front of your other squad members' fires, either.

To protect yourself, it's important to consider the difference between *cover* and *concealment*. Cover will afford you a degree of protection from enemy direct or indirect fire. Depending on the type of cover, cover can also provide concealment from enemy observation. Concealment means the enemy can't see you, but concealment doesn't protect you from enemy direct or indirect fire.

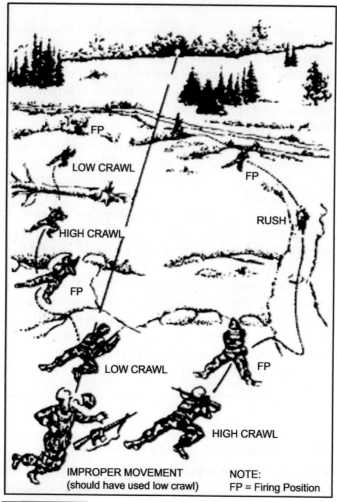

Figure 3.1 Three Types of Movement

There are varying degrees of cover and concealment. Tall grass or dense vegetation will help conceal your movement, but will not stop a bullet or shrapnel from direct or indirect fire. Getting behind a tree or a wall would help stop bullets, and may improve your concealment, but will not afford much protection from indirect fire. Occupying a position inside a building will improve your cover from direct and indirect fire and may offer better concealment from enemy observation.

So now you have identified your next covered position, and you know what route you want that provides the best cover or concealment. You have three options for movement: the *high crawl*, the *low crawl*, and the *three-second-rush*. You choose which one to use depending on the conditions you face—such as the terrain and the likelihood of enemy contact—or, if you are already receiving enemy fire, on the enemy fire's degree of accuracy. Features such as a gully, ditch, ravine, or wall can provide cover and concealment when you use the low or high crawl. Features such as hedgerows or lines of thick vegetation offer concealment only when you use the low or high crawl. (Remember that high grass or weeds only partially conceal you, since the movement of the grass as you crawl could give away your position.) Large trees, rocks, stumps, folds or creases in the ground, or vehicle hulks can give you cover and concealment in a temporary position.

If the enemy fire you are receiving is from a great distance or is inaccurate fire, it may be best for you to move quickly out of the enemy's line of fire by conducting three-second rushes. If the enemy's fire is close, and somewhat accurate or effective, you may need to high crawl out of the enemy's fires or to a covered position. If you are receiving close, accurate, or effective enemy fire, then in order to survive, you must give the enemy the smallest possible target by low crawling to the nearest cover.

An exception to this would be if you were the target of a close ambush. In this case, you would immediately return fire and assault through the ambush in order to get out of the kill zone and survive. This technique is known as a *battle drill*. You will learn more about battle drills later in ROTC.

The Low Crawl

The low crawl offers you the greatest protection with the slowest movement. Use the low crawl when you do *not* have to move quickly and you have less than a vertical foot of cover and concealment (or when the enemy has good visibility).
With the low crawl, you hug the ground:

1. Keep your body as flat as possible.
2. Grab the upper sling swivel of your weapon and let the weapon trail behind you (see Figure 3.2). The hand guard will rest on your forearm and the butt of weapon will drag on the ground. Keep the muzzle off the ground.
3. Push both arms forward and pull your right leg forward. Move forward by pulling with your arms and pushing with your right leg. Continue to push, pull, and move. Switch legs as you get tired. *Stay low.*

The High Crawl

Use the high crawl when you have to move quickly and your route offers cover and concealment (or when poor visibility limits enemy observation).
The high crawl is a modified version of crawling on your arms and legs:

1. Keep your torso off the ground and rest your weight on your forearms and lower legs—or your elbows and your knees.
2. Cradle your weapon in your arms and keep the muzzle off the ground (see Figure 3.3).
3. Keep your knees behind your buttocks so your buttocks stay low.
4. Move forward on your right elbow and left knee, then follow with your left elbow and right knee.

Figure 3.2 The Low Crawl

Figure 3.3 The High Crawl

The Three-Second Rush

The three-second rush—as the name implies—offers you the fastest movement with the least protection. *You will be exposed.* Use the rush when you have no cover or concealment, and breaks in enemy fire allow you to expose yourself *briefly.*

1. Roll or crawl away from your fighting position.
2. Push up with your arms. Spring to your feet. Carry your weapon at a modified position of port arms. Be ready to fire—or return fire—on the run.
3. *Run* to your next position. Run a short distance. Keep your exposure time to no more than three seconds. Do not let the enemy fire on you. Speed and surprise are your best friends.
4. Just before you hit the ground, plant both feet and fall forward. As you fall forward, slide your hand to the heel of the butt of your weapon, and use the butt of your weapon to break your fall.
5. Take up a good prone firing position and cover your buddy's movement.

In the last section, you learned that you work with a buddy on your fire team. Always move as a team. Cover one another. Never move without your buddy covering your movement. Never let your buddy move without you covering his or her movement with your weapon.

Figure 3.4 The Three-Second Rush

Communicate with your buddy. Make sure you and your buddy understand who moves when and where and when to provide covering fires to protect each other's movement. More important, in order to prevent fratricide (your buddy accidentally shooting you or vice versa), you and your battle buddy must also communicate when you will cease your covering fire.

The Two Fire Team Movement Formations

Fire teams, squads, and platoons use movement formations because:

- They allow the leader to maintain control over the Soldiers
- They allow the Soldiers to protect each other
- They allow the Soldiers to react flexibly when making contact with the enemy
- They make the best use of the team, squad, or platoon's firepower.

In both of the following fire team formations, the fire team leader moves at the front of the formation. This allows the fire team leader to lead by example, allows each fire team member to see the leader, and allows the fire team leader to fire and maneuver the fire team by using hand and arm signals.

Fire teams have two options for movement formations: the wedge formation and the file formation.

The Wedge Formation

The wedge is the basic fire-team formation. The Soldiers are spaced about 10 meters apart, depending upon the terrain. The team leader moves at the point of the wedge. Behind the leader and to his or her sides are the automatic rifleman and the grenadier. The rifleman trails the automatic rifleman or the grenadier in the wedge (see Figure 3.5). If the fire team is moving independently, or is the last element in part of a squad or platoon movement,

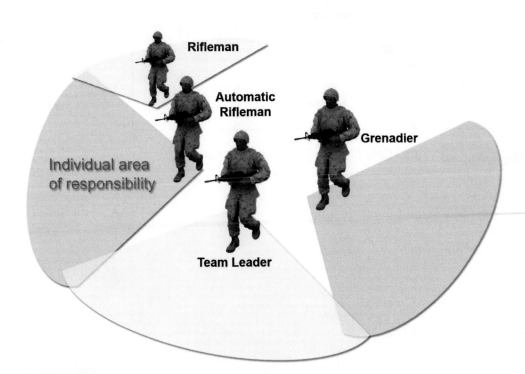

Figure 3.5 The Wedge Formation

the rifleman will occupy the center rear of the wedge formation and the wedge will resemble a diamond shape (see Figure 3.6).

The fire team leader adjusts the distance between Soldiers based on the terrain and the chance of enemy contact. If the terrain becomes restrictive, or if enemy contact is not likely, the fire team leader will contract, or collapse, the wedge formation by closing up the distance and dispersion between Soldiers. This allows easier command and control over the fire team. In severely restricted terrain or very low visibility, the fire team leader may collapse the wedge to the point where the wedge looks like a single-file formation. If the terrain is more open, or if enemy contact is likely or expected, the fire team leader expands the wedge formation by increasing the distance between Soldiers. This increases the difficulty of command and control, but also increases the protection and security of the fire team from enemy contact. In all cases, the fire team leader modifies the formation by reducing or increasing the interval—while still allowing each team member to see the fire team leader and the fire team leader to see the squad leader.

The wedge is easy to control, is flexible, provides good security, and allows the team members to fire immediately in all directions.

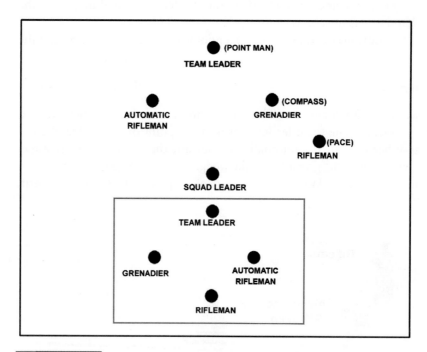

Figure 3.6 Squad Column With Fire Teams in Wedge

Critical Thinking

Why is it so crucial that team members be able to see one another? Why is control so important?

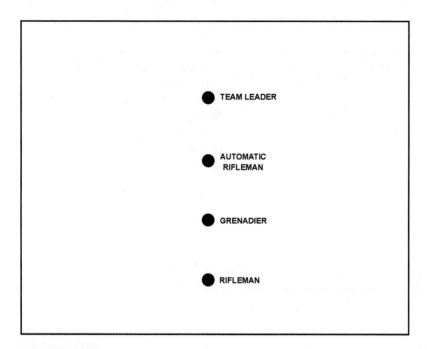

Figure 3.7 The File Formation

The File Formation

If terrain or visibility prevents the team members from using the wedge, they use the file formation. The team leader walks at the point, followed by the automatic rifleman, the grenadier, and the rifleman. They walk about 10 meters apart.

While the file provides the team leader greater control than the wedge does, it is less flexible, less secure, and prevents the team from firing to the front and rear.

TABLE 3.1	A Comparison of the Wedge and File Formations

CHARACTERISTICS					
Movement Formation	When Normally Used	Control	Flexibility	Fire Capabilities/ Restrictions	Security
Fire Team Wedge	Basic fire team formation	Easy	Good	Allows immediate fires in all directions	Good
Fire Team File	Close terrain, dense vegetation, limited visibility conditions	Easiest	Less flexible than the wedge	Allows immediate fires to the flanks, masks fires to the front and rear.	Least

Critical Thinking

Using Table 3.1, compare and contrast the characteristics of the fire team wedge and file. Why is the file easier to control than the wedge? How is it that the file is easier to control than the wedge, but is less flexible than the wedge? Consider fire capabilities and dispersion. Why does the wedge afford better security than the file?

CONCLUSION

An Army adage dictates, "The more you sweat in training, the less you bleed in battle." In the opening vignette, Holley was reminded, "just remember what you learned in basic training—when the lead's flying, get your butt down."

The Army has developed individual movement techniques and movement formations through years of experience in infantry tactics during combat. Just as Holley discovered, if you are to survive in combat, your training must become natural, instinctive, and automatic. The better you learn these techniques, the better your chances of surviving and prevailing in combat. You learned about individual movement techniques and moving as a buddy team in this section. In later ROTC courses, you will learn how small units use movement techniques and formations to survive on the battlefield.

Learning Assessment

1. Describe the advantages and disadvantages of the low crawl.
2. Describe the advantages and disadvantages of the high crawl.
3. Describe the advantages and disadvantages of the three-second rush.
4. Describe the advantages and disadvantages of the wedge formation.
5. Describe the advantages and disadvantages of the file formation.
6. Which Soldier takes point in a fire team formation?

Key Words

maneuver
cover
concealment

References

Field Manual 1-02, *Operational Terms and Graphics*. 21 September 2004.
Field Manual 7-8, *Infantry Rifle Platoon and Squad*. Change 1. 1 March 2001.
Hackworth, D. (2002). *Steel My Soldiers' Hearts*. New York: Simon and Schuster.
STP-21-1-SMCT, *Soldier's Manual of Common Tasks*. 31 August 2003.

INTRODUCTION TO MAP READING

Key Points

1 Marginal Information

2 Topographic Symbols

3 Terrain Features

4 Determining Four- and Six-Digit Grid Coordinates

Introduction

In Section 1, you learned how to navigate using information from a civilian-style map and a compass. In doing so, you learned that in order to navigate accurately, the map is one of your most important pieces of equipment. In this section, you will examine a military map, study its parts, and learn more about its uses. To be safe in a battle zone, you must know how to read a map, plot your location, and move in the right direction. If you can't navigate correctly, you risk getting lost—or worse, stumbling into dangerous territory. Consider the experience of MAJ Robert K. Wright, Jr., historian for XVIII Airborne Corps. MAJ Wright accompanied the Corps in Operation Just Cause, the American liberation of Panama in 1989.

Lost in Panama

I had one last interview to do on [January] 13th[, 1990].... So I went over and got that interview; they were off at a different location, so I'd gotten a driver to take me over, and I got one of their drivers to take me back to Fort Clayton, to the battalion headquarters. And I'd really gotten to know that battalion ... very, very well while I was down there. So I asked the S-3 could he get me a ride to the airport. So he gave me an NCO and a driver and a 'Hummer' [HMMWV; M-998-series High Mobility Multi-Wheeled Vehicle] and said "Sure, just take the Doc out."

So we swung by, picked up my gear. I cleared post. And off we went. And we're driving and driving and driving, and I know it isn't that far. Plus, we're going through the jungle. We're going up a paved highway and everything, but passing traffic and whatnot, which is taking forever. And then we went past this one area and I recognized it from aerial recon that I had done in the helicopter photography missions—this was Cerro Tigre, the PDF [Panamanian Defense Force] supply depot. Which was about 120 degrees in the wrong direction from the airport.

So at that point I casually inquired of the driver "Do you know where we're going?" And he said, "Why no, sir, I thought you knew where we're going." And I turned around and looked at the NCO, and he said "Don't look at me, I don't have a map either." So I said "Oh, O.K., well, hang a right and we'll keep going until we find the ocean or something and we get oriented." And we literally wandered around.

And I remembered thinking at the time, yeah, I've got seven rounds in my .45 . . . . So here we are, traveling through the countryside and had . . . I mean, we were out in the boonies. And had there been a disgruntled PDF guy still running around loose, it was me and my seven rounds from the .45, and that's all we had to protect us (Department of the Army, XVIII Airborne Corps, 1990).

Critical Thinking

What mistake did MAJ Wright, his driver, and the NCO make? Who was responsible for the mistake: MAJ Wright, the driver, or the NCO?

Marginal Information

The Army defines a map as "a graphic representation of a portion of the earth's surface drawn to scale, as seen from above (FM 3-25.26)."

Because the map is a *graphic* representation, you'll need a written explanation of the graphic elements. You'll find that explanation in the margins of the map: the *marginal information*. (Chapter 3 of FM 3-25.26 explains all the marginal information in detail.)

Figure 4.1 Scaled Representation of the Earth's Surface

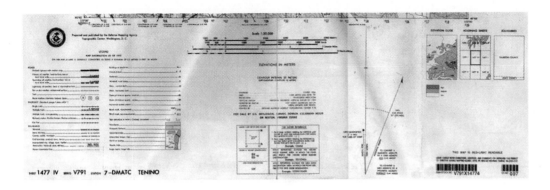

Figure 4.2 The Bottom of a Map

The map **legend** identifies the symbols used to depict the prominent natural and man-made objects that exist on the ground. These symbols are not the same on every map, especially foreign maps. Check the legend to avoid making serious mistakes. The legend from the bottom of the map in Figure 4.2 is shown enlarged in Figure 4.3.

The *sheet name and number* provide the title and the reference number for the map. Maps usually take their sheet names from the largest settlement or natural feature on the map. For example, the "Tenino Map" includes the community of Tenino, Washington.

The sheet number is in bold print in the upper right and lower left areas of the margin (Figures 4.2 and 4.3). At the lower right margin on the map is a diagram that shows adjoining map sheets. Your map sheet will always be depicted in the center of this diagram. You will learn later in your military studies how to link adjoining map sheets to operational overlays, operation orders, and operation plans.

legend

the section on a map that contains the symbols you need to read the map

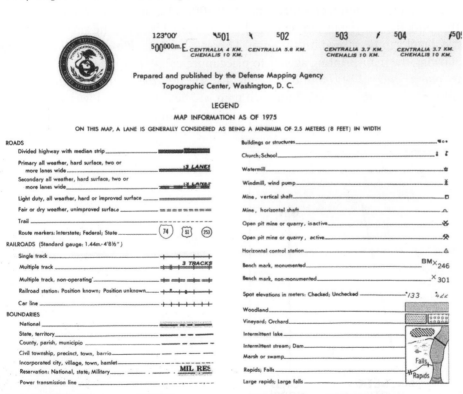

Figure 4.3 Map Legend

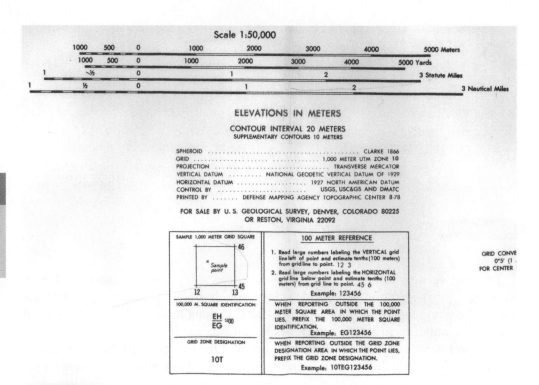

Figure 4.4 Map Scale, Contour Interval, and Grid Reference Box

scale

the ratio of the map distance to the corresponding distance on the earth's surface

The **scale** gives you the ratio of the distance on the map to the distance on the ground. For example, a scale of 1:50,000 (Figure 4.4) indicates that one unit of measure on the map equals one unit of measure on the ground. In other words, one inch on the map equals 50,000 inches on the ground, or approximately 8/10ths of a mile or 1.27 kilometers. The larger the ratio, the less detail can be placed on the map. Likewise, the smaller the ratio, the more detail can be placed on the map. Therefore, a 1:25,000 map will have larger grids; allowing the map-printing agency to place more details onto the map.

The *contour interval*, also found in Figure 4.4, specifies the vertical distance between contour lines. The contour interval for each map will be listed in the lower center of the map margin. Make sure you note whether the interval is in meters or feet.

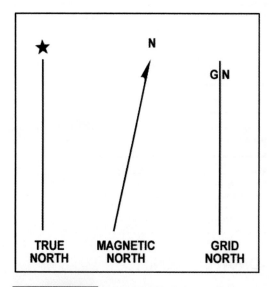

Figure 4.5 The Three Norths

The **declination** *diagram* depicts the three norths on your map: **true north**, **grid north**, and **magnetic north** (Figure 4.5). The declination diagram also lists the grid to magnetic angle (**G-M angle**) in degrees. The G-M angle is the angular difference between grid north and magnetic north. The closer to the poles you go, the greater this angle becomes. Understanding the G-M angle is critical. In order to use a grid map and compass to navigate, you must know how to convert a magnetic azimuth from your compass to a grid azimuth onto your map (or vice versa). You'll learn more about azimuths in the next section.

The *adjoining sheets diagram* tells you the sheet numbers of the adjoining sheets (Figure 4.6). You'll see a checkerboard-like display with the square in the center of the display representing the map you are reading. For example, if you need the map to the east of the map you're reading, look at the adjoining sheets diagram, identify the sheet number of the adjoining map, and request the map.

The *notes* tell you—among other things—the references the mapmakers have used in determining vertical and horizontal distances and the agencies responsible for the map information.

Mapmakers have divided the world into 60 grids and given those grids short letter-and-number (alpha-numeric) designators called *grid zone designators*. The grid zone designator for your map sheet is located at the bottom center of the map inside the *grid reference* box (Figure 4.4). The information in this box gives you the grid zone designation and the 100,000-meter square identification for your map sheet. You need to know the grid zone designator and the 100,000-meter square identification in order to convey information to others about your location or accurately call for indirect fire or close air support.

declination

the angular relationships between grid north and true north or magnetic north measured in degrees or mils east or west—a circle has 360 degrees or 6400 mils

true north

a line from any point on the earth's surface to the North Pole

magnetic north

the direction to the north magnetic pole, as indicated by the north—seeking needle of a magnetic instrument, such as a compass

grid north

the north that is established using the vertical grid lines on a map

G-M angle

the angular distance between grid north and magnetic north

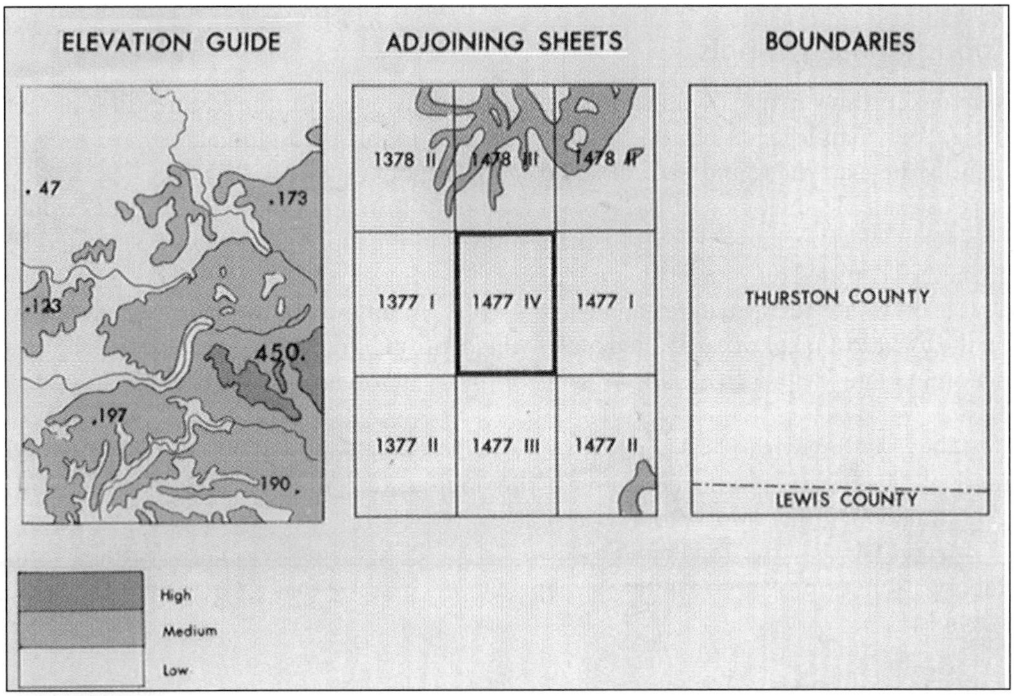

Figure 4.6 Adjoining Sheets

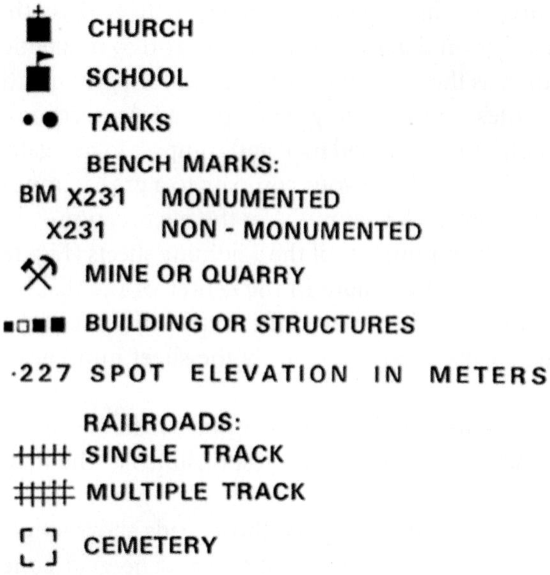
CHURCH
SCHOOL
TANKS
BENCH MARKS:
BM X231 MONUMENTED
X231 NON - MONUMENTED
MINE OR QUARRY
BUILDING OR STRUCTURES
·227 SPOT ELEVATION IN METERS
RAILROADS:
SINGLE TRACK
MULTIPLE TRACK
CEMETERY

Figure 4.7 Topographic Symbols

topographic symbols

symbols on a map that represent natural and man-made features found on the earth's surface

*The word **topography** comes from two Greek words: **topo**, meaning "place or region," and **graphos**, meaning "to write." So a topographic symbol or a topographic map is a detailed and precise description of a place or region.*

In order to maintain operational security and your night vision, you will use a red-lens flashlight when conducting operations during limited visibility. Older maps printed major roads and built up areas in a red ink that "disappeared" in red-filtered light. Imagine the difficulty of navigating at night when your major roads and built up areas disappeared after you turned on your red-lens flashlight! Mapmakers have recognized the problem and now print military maps with a red-brown ink color that is red-light readable.

Topographic Symbols

Military maps show various man-made and natural features using **topographic symbols** and different colors.

Topographic Symbols

Mapmakers draw maps so you can visualize the landscape with the features in the right place. Your map legend defines the topographic symbols the mapmakers have used to identify the man-made and natural features on the map (Figure 4.7).

For example, the topographic symbol used on your map to represent a school would be a small, black rectangle with a pennant drawn on the top. Another example would be a vineyard depicted on your map as a series of close tiny green dots. The legend may show a place of worship as a small rectangle with a cross, an upward arrow, or a crescent drawn on the top. Most maps of the United States will identify churches—no matter the religious denomination—with a cross. This practice will vary in foreign areas. Check the legend to be sure.

The legend may show a cemetery as a small rectangle drawn with dotted lines and marked "Cemetery." In foreign areas, the mapmakers may indicate the religious denomination, if that information is available.

Army FM 21-31, *Topographic Symbols*, describes the symbols, features, and abbreviations approved for military maps. Do *not* assume that all maps use the same symbols.

Colors

Imagine the difficulty of using a map printed only in black and white. Roads and rivers would look the same—probably with disastrous consequences. As early as the 15th century, mapmakers were coloring their maps. The use of color has become standardized, but check the legend to be sure.

1. *Black* indicates cultural (man-made) features such as buildings, railroads, and roads.

2. *Red and brown* combinations identify cultural features (such as major roads), relief features, and contour lines on red-light readable maps.

3. *Blue* identifies water: lakes, swamps, rivers, and coastal waters.

4. *Green* identifies vegetation such as woods, orchards, and vineyards.

5. *Brown* identifies cultivated land on red-light readable maps. On older maps, brown represents relief features and elevation such as contours.

6. *Red* was used on older maps to mark populated areas, main roads, and boundaries.

7. *Other colors* may show special information. Check the legend.

> *Be aware of how the seasons and climate may affect the presence or depth of intermittent water sources. Maps will display intermittent water sources as blue, but water may not actually be present at the time you are navigating.*

Terrain Features

As you look at the land around you, you will notice different **terrain features**: the hills, valleys, and other features on the ground. Maps represent these features in specific ways.

The Army divides terrain features into three groups: major, minor, and supplementary terrain features.

Major terrain features include *hills, saddles, valleys, ridges,* and *depressions.*

> **terrain features**
>
> *characteristics of the land, such as hills, ridges, valleys, saddles, depressions, and so forth*

a. A *hill* is an area of high ground. If you stand on a hilltop, the ground slopes away from you in all directions. A map represents a hill with contour lines forming concentric circles. The inside of the smallest circle is the hilltop (Figure 4.8).

b. A *saddle* is a dip or a low point between two areas of higher ground. If you stand in a saddle, you have high ground in two opposite directions and lower ground in the other two directions. The contour lines on a map representing a saddle are shaped like an hourglass (Figure 4.9).

c. A *valley* is a groove in the land, usually formed by a stream or a river. A valley usually begins with high ground on three sides and has a course of running water through it. If you stand in a valley, you will have higher ground in three directions and lower ground in one direction. Depending on the size of the valley and where you are standing, you may not see the higher ground in the third direction, but the stream or the river will flow from higher to lower ground (Figure 4.10).

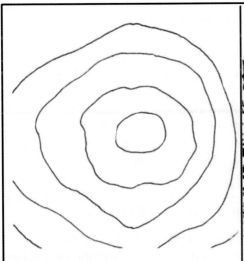

HILL

Figure 4.8 Hill

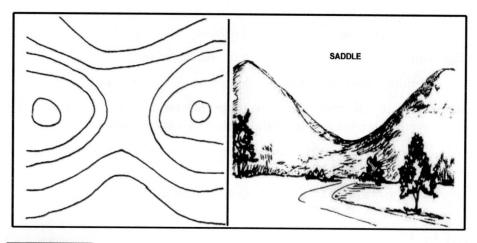

Figure 4.9 Saddle

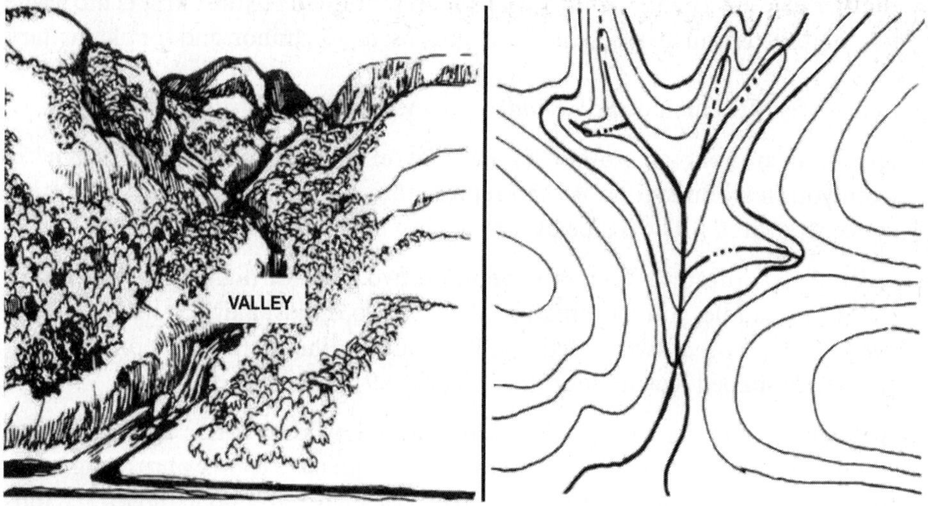

To the untrained eye, the contour lines of a valley and a draw (Figures 4.10 and 4.13) look very similar on a map. From a military consideration, a valley will afford a degree of maneuver room for troops and equipment, whereas a draw may only accommodate a small maneuver element, such as a squad or platoon.

Figure 4.10 Valley

A map represents a valley with U-shaped or wide V-shaped contour lines. Look at the contour lines to determine the direction the stream or the river is flowing. The closed end of the contour lines (the U or the V) points upstream and toward higher ground.

Not too long ago, the military term for a ridge was **ridgeline**. Many older Soldiers will still refer to a ridge as a ridgeline, as ridges will generally join a series of hills along a line.

d. A *ridge* is a sloping line of high ground. Think of a ridge as the high ground that runs along a hill. A series of hills connected forms a ridgeline. If you stand on the centerline of a ridge, you will normally have low ground in three directions and high ground in one direction. If you cross a ridge, you will climb to the crest and descend to the base. A map represents a ridge with U-shaped or V-shaped contour lines, but, unlike a valley, the closed end of the contour lines point to lower ground. A ridge can be easily confused with a spur (see below). The difference is that a spur will generally run perpendicular to a ridge or ridgeline, while a ridge will run directly off a hill or a series of hills (Figure 4.11).

e. A *depression* is a sinkhole, a pit, or a low point in the ground. Think of a depression as an upside-down hill. If you stand in the center of a depression, you will have higher ground in all directions. A map represents a depression with contour lines forming concentric circles; tick marks point to the lower ground (Figure 4.12).

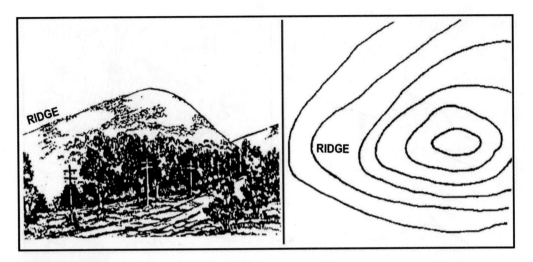

Figure 4.11   Ridge

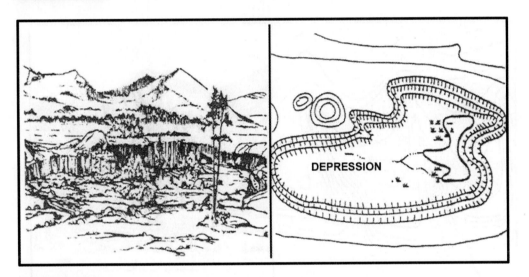

Figure 4.12   Depression

Minor terrain features include draws, spurs, and cliffs (Figures 4.13 through 4.15).

    a. A *draw* is a small valley. A draw has essentially no level ground and little or no maneuver room. If you are standing in a draw, the ground slopes upward in three directions and downward in the other direction. You could consider a draw to be the initial formation of a valley. A valley will usually have many draws feeding into the valley with streams or intermittent streams feeding into the body of water flowing through the valley. On a map, the contour lines depicting a draw are sharply V-shaped, pointing to higher ground. In most cases, a draw will be situated to the left or right of a spur or lying between two spurs (Figure 4.13).

    b. A *spur* is a short ridge. The ground will slope downward in three directions and upward in one direction. On a map, the contour lines depicting a spur are U-shaped pointing away from higher ground. In most cases, a spur will have draws to the left or right, or a spur is situated between two draws (Figure 4.14).

    c. A *cliff* is a vertical or near-vertical feature. On a map, the contour lines for cliffs are nearly touching or the contour lines come together to form one contour line depicting the edge of the cliff. Newer maps may also depict a cliff with the

Figure 4.13 Draw

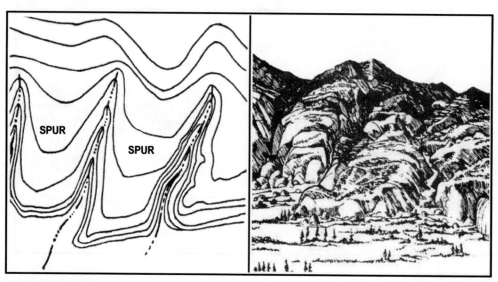

Figure 4.14 Spur

Figure 4.15 Cliff

same type tick marks used in depicting a depression, with the tick lines facing downward representing the vertical face of the cliff (Figure 4.15).

Supplementary terrain features include cuts and fills (Figure 4.16).

a. A *cut* is a man-made feature that cuts through raised ground. You may see a cut on a map forming a level bed for a road or railroad track.

b. A *fill* is a man-made feature that fills a low area. Again, you may see a fill on a map forming a level bed for a road or railroad track.

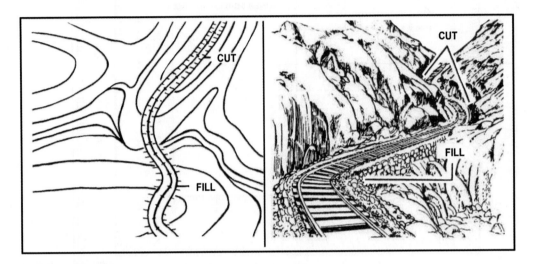

Figure 4.16 Cut and Fill

Using Four- and Six-Digit Grid Coordinates to Determine a Location

Grid coordinates are very important to the daily life of the Soldier. Soldiers use grid coordinates to find locations or convey locations on maps to others. They use grid coordinates to navigate, report enemy activity, request medical evacuation, or call for additional supplies and ammunition. Soldiers use grid coordinates to request indirect fire support from field artillery and naval gunfire. They also use grid coordinates to request close air support from fixed- and rotary-wing aircraft. As you read earlier, you'll find the grid reference box at the bottom center of the map. This gives you the grid zone designation and the 100,000-meter square identification for your map sheet. With more-exact grid coordinates you can more precisely plot or convey a location on the map. An important tool for doing so is your protractor.

Protractor

A **protractor** is a tool for working with maps. Protractors have an index mark in the center and divide a 360-degree circle into units of angular measure that are marked on two scales (degrees and mils) along the outside edge. The index mark is the center of the protractor circle, from which you measure all directions.

The Army protractor is Graphic Training Aid (GTA) 5-2-12, 1981 (Figure 4.17). It has four major parts:

1. A cross-hair in the middle, which you use to reference the north-south and east-west grid lines on a map

grid coordinates

letter and number designations that allow you to locate a point on a map

Think about the coordinates on the map at a multi-story shopping mall. The directory might tell you that the store is at F23. You examine the map and find that Section F is located on the third floor. Store 23 is in a side corridor at a right angle to the main shopping corridor, just before you reach the food court. Once you've located the store on the map, you find the nearest escalators and you're on your way. Using grid coordinates on a military map is quite similar.

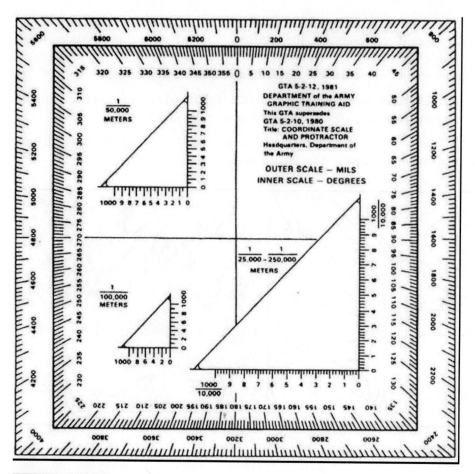

Figure 4.17 Army Protractor (GTA 5-2-12, 1981)

2. Three scales-1:100,000; 1:50,000; and 1:25,000

3. An inner scale of 360 degrees, which you use to plot azimuths (You'll learn more about azimuths in the next section.)

4. An outer, mils scale. (There are 6400 mils in a circle. You'll learn to use this scale for indirect fire.)

Using Four-Digit Grid Coordinates to Determine a Location

Earlier in this section you learned that mapmakers break down the earth's surface into 60 grid zone designators. The grid zone designator for your Tenino map, for example, is 10T. Each of these grid zone designators covers very large areas of the earth's surface. Because grid zones are not manageable in size when navigating, the mapmakers further break down each grid zone into 100,000-meter squares to make the grid zones more manageable. This means that the distance between each grid line is 100,000 meters.

For example, the area on your Tenino map covers portions of two 100,000-meter squares, and their identification is EH and EG. Unless you are flying, you will never need to navigate over an area as large as a 100,000-meter square. So mapmakers break down the earth's surface within the 100,000-meter squares into 10,000-meter squares (Figure 4.18) and then into even smaller, 1,000-meter squares (Figure 4.19) and number them beginning with 00 and ending in 99 (see Figure 4.4). Between each number, 01 and 02 for example, the distance is 1,000 meters.

Now imagine you are behind enemy lines and you are in satellite radio contact with your rescue aircraft, which is in another part of the world. You cannot simply give your

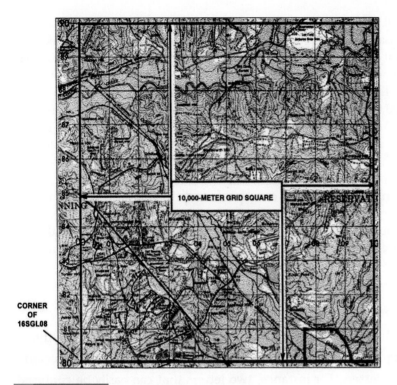

Figure 4.18   10,000-Meter Grid Square

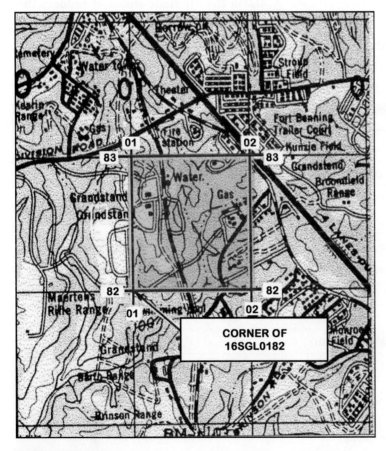

Figure 4.19   Four-Digit or 1,000-Meter Grid Square

rescuers the four-digit grid to your location, because every one of the 60 grid zones has thousands upon thousands of similar four-digit grids. You radio the aircraft, "Rescue 6 this is Lost Sheep 3—I am in grid zone Sixteen-Sierra." With this information, your rescuers can now narrow down your location on the earth to one of the 60 grid zones around the world. They begin to fly towards your area of the world, but need to narrow their search. You radio in your 100,000-meter square identification, "Rescue 6, I am at Sixteen-Sierra, Gold Lima." Your rescuers now know where you are within a 100,000-meter square. This is still too large of an area to search, so they ask you for more detailed coordinates. You radio back, "Rescue 6, I am at Sixteen-Sierra, Golf Lima, Zero One, Eight Two (Figure 4.19)." With this information. The pilots have to search only a 1,000-meter square, or one grid square on your map. Aided by your complete four-digit grid, your rescuers are able to spot your infrared emergency beacon and rescue you in a short period of time.

The pairs of numbers on the horizontal (east-west) and vertical (north-south) grid lines on your map are used to identify grid squares. Every set of grid coordinates will

---

The Phonetic Alphabet

The phonetic alphabet is used to spell out letters in place of just saying the letter itself. By using a word for each letter, there is less chance that the person listening will confuse letters. For instance, two letters that can easily be confused are "D" and "B." When a speaker uses the phonetic alphabet, a listener can easily distinguish between "*Delta*" and "*Bravo*." The phonetic alphabet is used primarily used in two-way radio communications. Using the phonetic alphabet reduces the effects of noise, weak signals, distorted audio, and radio operator accent. Maritime units, aircraft, amateur radio operators, and the military around the world use this system of spelling letters.

Letter	Pronunciation	Letter	Pronunciation
A	Alpha (AL fah)	N	November (no VEM ber)
B	Bravo (BRAH VOH)	O	Oscar (OSS cah)
C	Charlie (CHAR lee)	P	Papa (pah PAH)
D	Delta (DELL tah)	Q	Quebec (keh BECK)
E	Echo (ECK oh)	R	Romeo (ROW me oh)
F	Foxtrot (FOKS trot)	S	Sierra (see AIR rah)
G	Golf (GOLF)	T	Tango (TANG go)
H	Hotel (hoh TELL)	U	Uniform (YOU nee form)
I	India (IN dee ah)	V	Victor (VIK tah)
J	Juliet (JEW lee ETT)	W	Whiskey (WISS key)
K	Kilo (KEY loh)	X	X Ray (ECKS RAY)
L	Lima (LEE mah)	Y	Yankee (YANG key)
M	Mike (MIKE)	Z	Zulu (ZOO loo)

Note: Stress the syllables printed in capital letters.

have an even set of numbers. In a four-digit grid, the first half of the grid coordinate numbers represents the horizontal, "left-to-right" or "easting" reading. The second half of the grid coordinate numbers represents the vertical, "bottom-to-top" or "northing" reading. For example, grid coordinate 16GL0182 in Figure 4.19 would identify all of the area within the grid square to the right of line 01 and above line 82.

The *critical* rule is to read *right* and then *up*. Notice how the example reads *right* and then *up*: Grid square 0182 was to the *right* of line 01 and above—*up* from—line 82.

Using Six-Digit Grid Coordinates to Determine a Location

Submitting a four-digit grid location may be acceptable for large-scale operations or large-scale units. For example, a one-grid-square location might be sufficient for identifying the location of a brigade combat team forward operating base or a zone reconnaissance for a company-sized element. There are other situations, however, where your grid locations must be narrowed down in order to be more accurate than a 1,000-meter square For situations in which you need to be within a 100-meter square—such as calling for indirect fire or close air support, or calling for an emergency resupply or medical evacuation—you will need to know how to determine and plot six-digit grid coordinates.

Think back to the earlier search-and-rescue scenario. Imagine you are hunkered down in hiding because enemy forces are actively searching for you. It is crucial to your survival that your rescuers find you quickly. Rather than have them search an entire grid square for you, you radio your rescuers, "Rescue 6, this is Lost Sheep 3, I am at grid Sixteen-Sierra, Golf Lima, One-Four-Two, Eight-Four-Eight (Figures 4.20, 4.21 and 4.22). Rather than searching for an hour, your rescuers hover within 100 meters of your location within a matter of minutes.

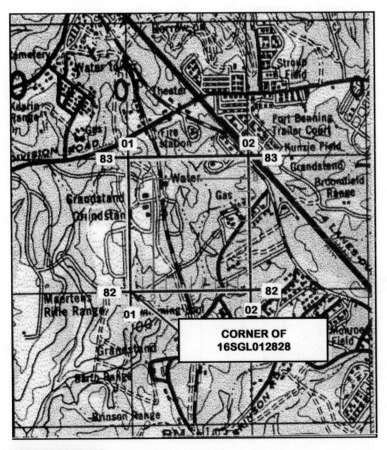

Figure 4.20 Six-Digit or 100-Meter Grid Square

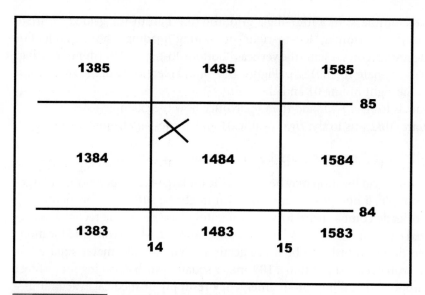

Figure 4.21 Plotting a Specific Point

Follow these five steps to identify a more specific location:

1. Make sure you are using the appropriate scale (check the scale in the map's marginal information) and make sure the scale is right side up.

2. Place the protractor scale with the zero-zero point at the lower left corner of the appropriate grid square.

3. Keep the horizontal line of the protractor's scale directly on top of the horizontal, left-to-right, or "easting" grid line, and slide the protractor—and its scale—to the right until the left vertical line of the grid square touches the point on the protractor scale for the coordinate you want.

4. Read up the vertical scale until you reach the coordinate you want.

5. Mark the location.

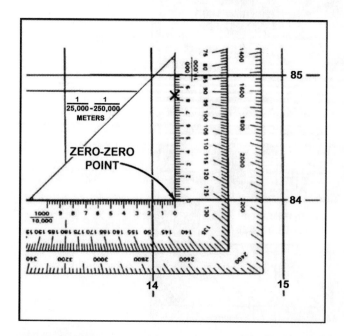

Figure 4.22 Read Right, Then Up: Six-Digit Grid Coordinate 16SGL142848

Make sure the horizontal line of the protractor's scale is lined up with the horizontal, left-to-right, or "easting" grid line, and the vertical line of the scale is parallel with the vertical, bottom-to-top, "northing" grid line.

Remember to write coordinates as one continuous number without spaces, parentheses, dashes, or decimal points. Write them as an even set of numbers so that whoever uses your coordinates knows where to make the split between the *right* and *up* readings.

Be *very* careful not to misidentify or transpose the grid numbers. Double- and triple-check the numbers. Ask someone else to review your numbers. If you send wrong or transposed numbers, the rescuers may not find you or the artillery rounds will not fall where you want them to—they may fall on your position.

Critical Thinking

Discuss some consequences (other than those discussed in the text) of misidentifying or transposing grid coordinates. Think about how far off your grid coordinates will be if you just make one numerical error.

How could you make sure your grid coordinates are accurate in a pressure situation?

CONCLUSION

MAJ Wright, his driver, and an NCO wound up "out in the boonies" because none of them had looked at a map before leaving for the airport. Knowing how to read a map and plot coordinates are essential military skills. In this section, you learned how to determine and plot a grid coordinate with 100-meter square accuracy. In later sections, you will learn how to plot and determine grid coordinates within a 10-meter square, by using eight- and 10-digit grid coordinates. During any mission you must always know where you are and where you are going. If you are to be a credible leader, your Soldiers must be confident that you are proficient in map reading and land navigation. Take the time now to gain and polish your map-reading skills. Not only are they important to your Army career—they can save your life or your Soldiers' lives in combat.

Learning Assessment

1. Describe and explain in your own words the five major terrain features of hill, valley, saddle, ridge, and depression.

2. Explain the difference between a draw and a valley; and between a spur and a ridge.

3. Define man-made and natural objects depicted on a military map by topographical symbols.

4. Which is more precise—a four-digit or six-digit grid location?

5. Explain how to determine a four-digit grid location of an object on a military map.

6. Explain how to determine a six-digit grid location of an object on a military map.

7. Describe how to identify a four-digit grid coordinate on a military map.

8. Describe how to identify a six-digit grid coordinate on a military map.

Key Words

legend

scale

declination

true north

magnetic north

grid north

G-M angle

topographic symbols

terrain features

grid coordinates

protractor

References

Department of the Army. XVIII Airborne Corps. (6 April 1990). Joint Task Force South in Operation Just Cause. *Oral History Interview JCIT 046*. Fort Bragg, NC. Retrieved 8 July 2005 from http://www.army.mil/cmh-pg/documents/panama/jcit/JCIT46.htm

Field Manual 1-02, *Operational Terms and Graphics*. September 2004.

Field Manual 3-25.26, *Map Reading and Land Navigation*. 18 January 2005.

Field Manual 21-31, *Topographic Symbols*. Change 1. December 1968.

INTRODUCTION TO LAND NAVIGATION

Key Points

1 **Understanding Azimuths**

2 **Converting Azimuths**

3 **Determining Elevation**

4 **Calculating Distance on a Map**

Introduction

To accomplish your mission, you must be in the right place at the right time. Being in the right place requires you to navigate well. Knowing how to read a map is one thing—knowing how to use a map to navigate requires that you understand how to use azimuths, elevation, and map distance.

In the previous section, you learned how to identify and interpret topographic symbols, colors, contour lines, and marginal information found on a military map. You also learned about the military grid reference system and how to plot grid coordinates using a military map and protractor.

This section will expand your map-reading skills and introduce you to how the military navigates using a map, compass, and protractor. You will learn what an **azimuth** is and how to convert azimuths in order to navigate using a compass and map. You will also learn how to determine the elevation of the terrain by analyzing the contour lines and contour interval data from the marginal information on a military map. Lastly, you will learn to compute straight-line and road distance using the scale in the margin of the military map. Coupled with your learning from your orienteering and map reading lessons, you will have the basic knowledge to navigate from one point to another and arrive safely at your destination.

In the following vignette, Colonel John Zierdt, Jr., commander of the 1st Support Command during the first Gulf War, remembers how a group of Soldiers paid a serious price when they decided to rely on familiarity rather than put into practice basic map reading and land navigation skill required of all Soldiers.

Captured During Desert Storm

The driver had been on a particular route two or three times and thought he knew where he was going. Then instead of turning left, he kept going straight. They even saw the water on their right, which was a dead giveaway that they were going north rather than west. There were two HETs [heavy trucks] following each other. The guy, the one that was eventually captured, was in the lead vehicle, and stopped. And the guys behind him said, "You're going the wrong way and we need to turn around." He said, "I am not." He says, "I'm going straight. You can follow me or turn around if you want."

So, they kept going straight. The next thing you knew they were in the middle of a firefight. The second vehicle got turned around in time [and] got out of there; the [first] vehicle got stuck and didn't get turned around, and the two of them got captured (Department of the Army, XVIII Airborne Corps et al., 1991).

Critical Thinking

If the drivers of the two vehicles had looked at and oriented their maps, what might have told them they were headed in the wrong direction?

What would you have done if you were in the second vehicle? Would you have continued to follow the first vehicle after you decided it was going the wrong way?

What could you have said over the radio to the Soldiers in the first vehicle that may have triggered in their minds that they were, in fact, going in the wrong direction?

Understanding Azimuths

azimuth

the horizontal angle, measured clockwise by degrees or mils between a reference direction and the line to an observed or designated point—there are three base (reference) directions or azimuths: true, grid, and magnetic azimuth

Everything in land navigation begins with an **azimuth**. An azimuth is a horizontal angle measured clockwise by degrees or mils between a reference direction and a line to an observed or designated point. There are three base directions or azimuths: true, grid and magnetic.

The Army uses azimuths to express direction. Direction is determined from your start point, or where you are, outward toward your desired destination, or your intended target. Because you use north (0 or 360 degrees) as your base line, 270 degrees away from north will always be due west.

Think of yourself as standing in the middle of a Nebraska cornfield. You are facing north. The horizon stretches around you in a great 360-degree circle. If you travel an azimuth of zero degrees—or 360 degrees—or due north—you will wind up in Canada.

The terms **azimuth** and **direction** *are interchangeable.*

If you turn to your right and travel on an azimuth of 90 degrees—due east—you will wind up in the Atlantic Ocean, probably off the coast of New Jersey.

An azimuth of 180 degrees—due south—will take you into Mexico, and an azimuth of 270 degrees—due west—will take you to the Pacific, just off the coast of Northern California.

Determining a Grid Azimuth Using a Protractor

There are two ways you can determine an azimuth. You can use a map to determine a grid azimuth, or you can use a compass to determine a magnetic azimuth. Regardless of the technique, you will learn in this chapter how to convert a grid azimuth to a magnetic azimuth and a magnetic azimuth to a grid azimuth. You will first use a map and learn how to determine a grid azimuth. The steps in this process should be very familiar if you have ever taken a geometry class.

grid azimuth

the angle between grid north and a line drawn on the map

To begin, select a start point on the map. Mark it as point A. Identify an end point on your map. Mark it as point B. Using the edge of your protractor, draw a straight pencil line between points A and B. The line is your azimuth. Now you must determine the **grid azimuth** of that line—the angle between the line and grid north.

When you lay your protractor down on your map, make sure you place it right side up; verify this by checking to see that the writing on the protractor is not backwards. If your protractor is wrong side up, you will get grid azimuths that are 180 degrees off from

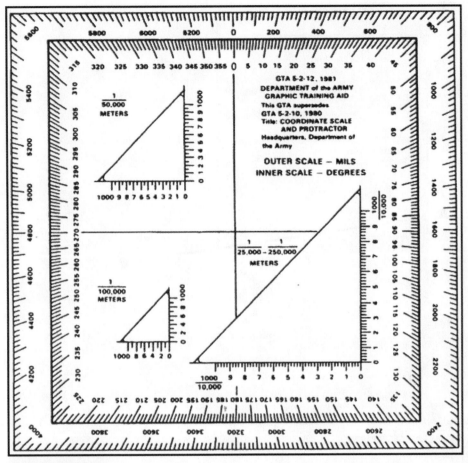

Figure 5.1 Army Protractor (GTA 5-2-12, 1981)

Although having the mils scale on the outside of the protractor may seem confusing now, don't get into the habit of cutting the mils scale off your protractor. Later in your military career, your military occupational specialty (MOS) may require you to state azimuths in mils as well as degrees.

the correct grid azimuth. Also, make sure the 0 or 360-degree mark of your protractor is toward the top (or north) of your map, and make sure the 90-degree mark is toward the right (or east) of your map. *If you place your protractor down incorrectly on your map, the grid azimuth that you determine will be a minimum of 90 degrees off and as much as 270 degrees off the actual azimuth.*

Follow these three steps to determine your grid azimuth from the arbitrary points A and B (Figure 5.2):

1. Place the index of your protractor (the place where the etched vertical line and the etched horizontal line meet) at the point where the line you drew on your map crosses a vertical, north-south grid line.

2. Keeping the index at this point, line up the 0-to-180-degree line, or base line, of the protractor on the vertical, north-south grid line.

3. Follow your line outward to the degree scale of your protractor. Read the value of the angle from the protractor. This is your grid azimuth from point A to point B expressed in degrees.

Next, you will plot an azimuth from a known point on a map. Imagine you receive an order to move from your current position in a given direction. Plotting the azimuth on your map will allow you to see the terrain and objects you will need to navigate through along the entire length of your azimuth. The steps are as follows:

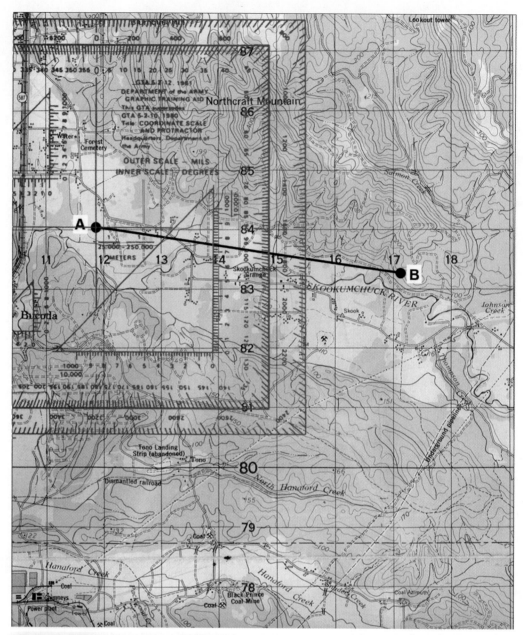

Figure 5.2 Measuring an Azimuth

This is the same method you will use to determine the grid azimuth between any two points on the map.

1. Place your protractor on the map with the index mark at the center of the known point and the base line parallel to a vertical, north-south grid line.

2. Using your pencil, make a small tick mark on the map at the edge of the protractor at the desired azimuth. Remember that your protractor will have degrees on the inner scale and mils on the outer scale. Ensure the tick mark on the map is beside the desired azimuth in degrees and not mils.

3. Lift and reposition the protractor so you can use its side as a straightedge. Draw a line connecting the known point and the tick mark on the map. This is your grid direction line—your azimuth.

Determining a Back Azimuth

A **back azimuth** is simply the opposite direction to your azimuth. A simple example is when you get on the interstate going north when you wanted to go south. At the next exit, you get off the interstate, turn around, and get back on the interstate going south. You just took a back azimuth, or in slang, you just "did a 180."

To compute a back azimuth from an azimuth, simply add or subtract 180 degrees to or from your original azimuth. Remember that a circle has 360 degrees, so if your azimuth is greater than 180 degrees, adding 180 degrees to determine your back azimuth will give you an azimuth that is more than 360 degrees. For example, if your azimuth were 200 degrees, adding 180 degrees would result in a back azimuth of 380 degrees, whereas subtracting 180 degrees would result in a back azimuth of 20 degrees. The back azimuth 380 degrees is obviously greater than the number of degrees in a circle-20 degrees greater. Sure, you could subtract 360 degrees from 380 degrees and still get the same correct back azimuth of 20 degrees. But this simply adds another step to the process. So, subtracting 180 degrees from azimuths greater than 180 degrees simplifies determining back azimuths.

back azimuth
the opposite direction of an azimuth—to obtain a back azimuth from an azimuth, add 180 degrees if the azimuth is 180 degrees or less, or subtract 180 degrees if the azimuth is 180 degrees or more

Determining a Magnetic Azimuth to an Object

A *magnetic azimuth* is an azimuth determined using magnetic instruments, such as a compass. The Army uses two types of compasses: the M2 compass and the lensatic compass. Soldiers use the M2 compass primarily for positioning indirect fire weapons such as mortars. The lensatic compass, pictured in Figure 5.3, is the compass the Army uses for land navigation.

To determine a magnetic azimuth using a compass:

1. Open your compass to its fullest so the cover forms a straightedge with the base. Move the lens (the rear sight) to the rearmost position. This allows the dial to float freely.

2. Place your thumb through the thumb loop, form a steady base with your third and fourth fingers, and extend your index finger along the side of the compass.

3. Place the thumb of your other hand between the lens (rear sight) and the bezel ring; extend your index finger along the remaining side of the compass, and your remaining fingers around the fingers of your other hand. Tuck your elbows into your sides. This will place the compass between your chin and your belt.

4. Turn your body toward the object that you wish to get an azimuth to, pointing your compass cover directly at the object.

5. Look down and read the azimuth from beneath the fixed black index line on the compass face.

Critical Thinking

1. Why is it important for you to understand how to determine a back azimuth?

2. When would you use a back azimuth?

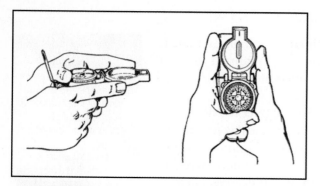

Figure 5.3 Centerhold Technique with a Lensactic Compass

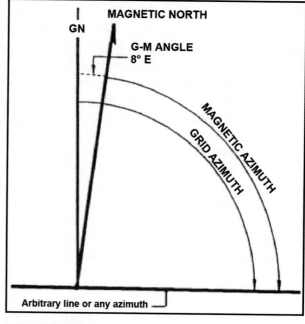

Figure 5.4 True, Magnetic, and Grid North

Shooting an Azimuth with a Compass

When you already know the magnetic azimuth that you want to navigate along, you follow the above steps, but reverse steps 4 and 5. You look down at the compass bezel and slowly turn your body until you see the azimuth you wish to take. Once you see your azimuth on the bezel, look up, and identify an easily recognizable object off in the distance that is in line with your azimuth. Once you have identified the object on your azimuth, you can put your compass away and move to that object. As long as you continue to move to your identified object, you will be on your desired azimuth. This method is known as "shooting an azimuth."

Converting Azimuths

Two problems complicate your easy use of a map and compass:

First, the surface of the earth is curved, while the surface of your map is flat. This creates problems between what your map shows as north (grid north) and what really is north (true north).

Second, the earth's magnetic pole is not the same as the earth's axis. This creates a difference between what your compass shows as north (magnetic north) and what really is north (true north).

Your map contains information to help you overcome these problems. As you read in Section 4, the **declination diagram** in your map's legend gives you the information you need to compensate for the differences—declination—between grid north, true north, and magnetic north.

The declination diagram (Figure 5.4) shows you the difference in angle between any of these norths. Since you will navigate with a magnetic compass and a grid map, your primary concern is the difference between grid north and magnetic north. The difference between grid north and magnetic north is known as the G-M angle (grid-magnetic angle).

declination diagram

the chart in the map legend that tells you the differences in angle between true north, grid north, and magnetic north

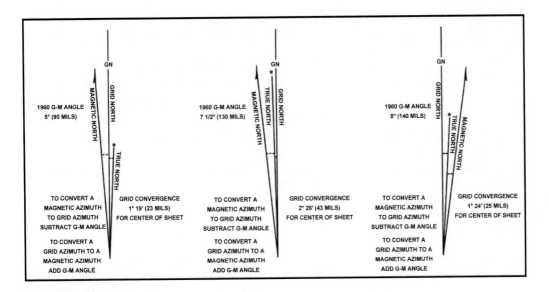

Figure 5.5 Map Declination Diagram

The G-M angle will be shown in the declination diagram and will be expressed in degrees. The G-M angle will either be to the west of grid north (westerly G-M angle) or to the east of grid north (easterly G-M angle). To reduce the confusion of converting easterly and westerly G-M angles from grid to magnetic or magnetic to grid, the mapmakers now include easy-to-understand instructions on the declination diagram so you can quickly convert azimuths without remembering formulas (Figure 5.5).

The three vectors that make up the declination diagram (true north, grid north, and magnetic north) are not drawn to scale. Use the written value for the G-M angle and do not try to measure the vectors to determine the G-M angle.

Most military maps will display the declination diagram in the lower margin. Some maps may not display the declination diagram and will only list the declination information as a note in the map margin.

Adjusting for the Grid-Magnetic (G-M) Angle

The G-M angle value is the size of the angle between grid north and magnetic north. You will see it as an arc, indicated by a dashed line, connecting the grid-north and magnetic-north vectors.

The G-M angle is important, because if you don't adjust for the G-M angle, your grid azimuth translated from your map to your compass will be wrong by the size of the angle and vice versa. For example, if your G-M angle is 8 degrees *and you don't adjust for that angle*, your grid or magnetic azimuth will be off by 8 degrees. The farther you move away from your start point on your incorrect azimuth, the farther off you will be from your objective. The angular error increases the farther you move. Not using the G-M angle when converting from a grid azimuth to a magnetic azimuth can cause you to miss your objective. Likewise, if you forget to use the G-M angle when you convert a magnetic azimuth to a grid azimuth, you will plot the wrong azimuth on your map. This could result in passing on incorrect information or calling in inaccurate indirect fire missions.

Look at the notes that accompany the G-M angle diagram (Figure 5.5). One note tells you how to convert your magnetic azimuth to a grid azimuth; another tells you how to convert your grid azimuth to a magnetic azimuth.

A typical note may read "To convert a magnetic azimuth to grid azimuth, subtract G-M angle." If you have a magnetic azimuth of 270 degrees, and the G-M angle is 8 degrees, your grid azimuth will be 262 degrees.

The conversion (whether to add or subtract) depends on whether your map has an easterly or westerly G-M angle. If your magnetic north is to the right (east) of the grid north, then your map has an easterly G-M angle. If your magnetic north is to the left (west) of the grid north, then your map has a westerly G-M angle.

You will learn more about azimuths and land navigation as you progress through ROTC. By the end of your MSL III year, you must master land navigation in order to succeed at the Leader Development and Assessment Course (LDAC), which you will attend at Fort Lewis, Washington, after your MSL III year.

Determining Elevation

You can determine the elevation of any location on your map without any special equipment using two things on your map that you learned about in the previous Map Reading section—contour lines and the contour interval. Before you can determine the elevation of any point on your map, you must first know the contour interval for the map you are using. As you read previously, you can find the contour interval in the margin of your map—usually in the middle of the lower margin. Recall that the contour interval is a measurement of the vertical distance between adjacent contour lines.

Refer to Figure 5.7 to learn how to determine the specific elevation of a point on a map:

1. Identify the contour interval and the unit of measure used (feet, meters, or yards) from your map's marginal information at Figure 5.6 (most military maps use meters).

    Using the map example at Figure 5.7, if you wanted to determine the elevation to point A, you would find the numbered index contour line nearest point A. In Figure 5.7, the closest numbered contour line to point A is the 500-meter contour interval.

    Determine if point A is above (higher in elevation) the 500-meter contour line, or if point A is below (lower in elevation) than the 500-meter line. Since point A lies between the 500-meter contour line and the 600-meter contour line, moving from the closest contour line (500-meter) to point A would be traveling uphill to a higher elevation.

2. Determine the elevation of point A by starting at the index contour line numbered 500 and counting the number of intermediate contour lines (the unmarked contour lines) to point A.

    Point A is on the second intermediate contour line above the 500-meter index contour line. Since the contour interval is 20 meters (Figure 5.6), each intermediate contour line crossed to get to point A adds 20 meters to the 500-meter index contour line. The elevation of point A is 540 meters.

3. Determine the elevation of point B by going to the nearest index contour line. In this case, it is the upper index contour line, numbered 600. Point B is located on the intermediate contour line immediately below the 600-meter index contour line.

    Therefore, point B is located at an elevation of 580 meters. Remember, if you are increasing elevation, add the contour interval to the nearest index contour line. If you are decreasing elevation, subtract the contour interval from the nearest index contour line.

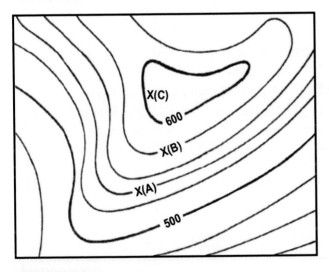

LEGEND

MAP INFORMATION AS OF 1975

ON THIS MAP, A LANE IS GENERALLY CONSIDERED AS BEING A MINIMUM OF 2.5 METERS (8 FEET) IN WIDTH

ROADS

Divided highway with median strip

Primary all weather, hard surface, two or more lanes wide — 3 LANES

Secondary all weather, hard surface, two or more lanes wide — 2 LANES

Light duty, all weather, hard or improved surface

Fair or dry weather, unimproved surface

Trail

Route markers: Interstate; Federal; State — 74 51 253

RAILROADS (Standard gauge: 1.44m.-4'8½")

Single track

Multiple track — 3 TRACKS

Multiple track, non-operating

Railroad station: Position known; Position unknown

Car line

BOUNDARIES

National

State, territory

County, parish, municipio

Civil township, precinct, town, barrio

Incorporated city, village, town, hamlet

Reservation: National, state; Military — MIL RES

Power transmission line

Buildings or structures

Church; School

Watermill

Windmill, wind pump

Mine, vertical shaft

Mine, horizontal shaft

Open pit mine or quarry, inactive

Open pit mine or quarry, active

Horizontal control station

Bench mark, monumented — BMx 246

Bench mark, non-monumented — x 301

Spot elevations in meters: Checked; Unchecked — •133

Woodland

Vineyard; Orchard

Intermittent lake

Intermittent stream; Dam

Marsh or swamp

Rapids; Falls — Falls

Large rapids; Large falls — Rapids

SHEET 1477 IV SERIES V791 EDITION 7–DMATC TENINO

Figure 5.6 Example of a Contour Interval Note

Figure 5.7 Points on Contour Lines

Critical Thinking

Why is it important for you to know how to determine elevation on a military map? Think about the azimuth you will plot on your map in order to travel from point A to point B. How will knowing elevation help you when navigating from point A to point B? Can knowing the elevation help you decide which azimuths or routes to take to your destination?

4. Estimate the elevation of the hilltop, point C, by adding one-half of the contour interval to the elevation of the last contour line. In this example, the last contour line before the hilltop is an index contour line numbered 600. Add one-half the contour interval, 10 meters, to the index contour line. The elevation of the hilltop would be 610 meters. You use the same process to estimate the elevation of a depression, except you subtract half of the contour interval to estimate the elevation at the bottom of the depression.

Calculating Distance on a Map

Now you know how to plot and determine azimuths on your map, and you understand how to determine elevation on your map or along your plotted azimuth. But how far is it from your start point to your destination? The marginal information on your map allows you to determine both straight-line distance and road distance. You can use the graphic scale—located in the lower center portion of the map margin—as a ruler to convert distances on the map to distances on the ground (Figure 5.8).

The graphic scale is divided into two parts. To the right of the zero, the scale is marked in full units of measure and is called the primary scale. To the left of the zero, the scale is divided into tenths and is called the extension scale.

Most maps have three or more graphic scales, each with a different unit of measure, such as meters, yards, statute miles, and nautical miles. When you use the graphic scale, be sure that you use the appropriate unit of measure.

Straight-Line Distance

To calculate the straight-line distance between two points on your map:

1. Lay a straight-edged piece of paper on the map so the edge of your paper touches both points and extends past them.

2. Make a tick mark on the edge of the paper at each point (Figure 5.9).

3. Then move your paper to the graphic bar scale, and use the scale to measure the distance between the two points. Note that you should align the tick mark on the right with a printed number in the primary scale so that the left tick mark falls within the extension scale (Figure 5.10).

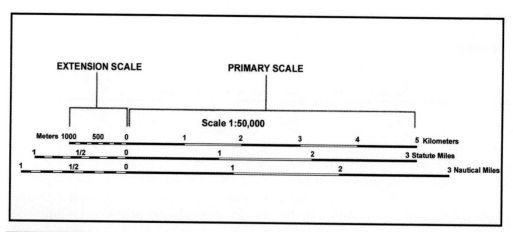

Figure 5.8 Using a Graphic (Bar) Scale

Figure 5.9 Transferring Map Distance to Paper Strip

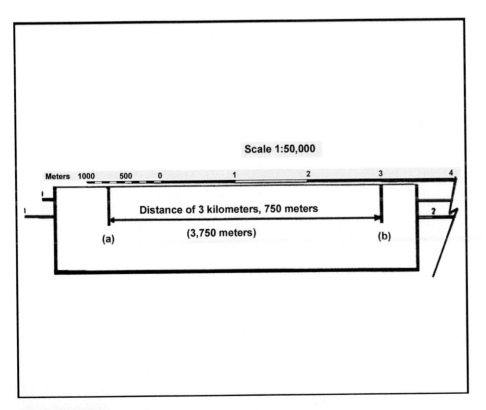

Figure 5.10 Measuring Straight-Line Map Distance

Curved-Line Distance

To measure the distance along a curved route, such as a road, trail, wateway, or other curved line:

Put a straight-edged piece of paper on your map with the edge next to your starting point. Place a tick mark on the paper and on your map.

Line up the straight edge of the paper with the straight portion of the curved route you are measuring. Make a tick mark on both map and paper when the edge of the paper leaves the straight portion of the line you're measuring. (See View A in Figure 5.11).

Pivot the paper until another straight portion of the curved line lines up with the edge of the paper. Continue in this manner until you have completed the distance you want to measure. (See View B in Figure 5.11. Notice the number of small ticks on the edge of the paperand that the last is labeled tick mark B.)

Move the paper to the graphic scale to determine the ground distance. The only tick marks you need to measure are tick marks A and B. (See View C in Figure 5.11.)

In order to maintain accuracy when measuring curved distance, it is important to keep the straight edge of your paper on the same side of the curve you are measuring. If you start off measuring a curved road on one side of that road, then keep your paper on that side of the road and mark all of your tick marks on that same side of the road. Do not cross over and start making tick marks on the opposite side of the road.

The more tick marks you make when measuring your curved route, the more accurate your final distance will be. This is especially true when measuring along curves.

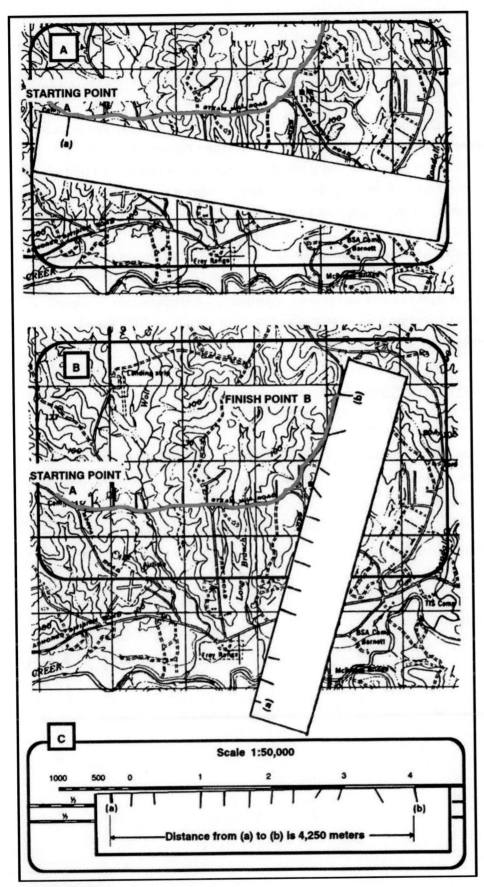

Figure 5.11 Measuring a Curved Line

CONCLUSION

You are a cadet now. In the not-too-distant future, you may be an Army second lieutenant leading a platoon. Perhaps, in the distant future, you will be a lieutenant colonel commanding a battalion, a colonel commanding a brigade, or even a major general commanding a division. Whatever your position and rank, you will always need to get your Soldiers from one point to another. If you can't do so, you endanger your mission and perhaps the lives of your Soldiers.

It's impossible to overemphasize the importance of map reading and land navigation. They are critical leadership skills. They are also perishable skills—they require constant practice and review, regardless of a Soldier's rank or experience. Start now to develop your expertise and work to keep your skills honed and at the ready.

Learning Assessment

1. What is an azimuth?
2. Explain how to determine a grid azimuth.
3. Explain how to determine a magnetic azimuth.
4. Explain the differences between the three norths.
5. Explain how to use the G-M angle to convert grid and magnetic azimuth.
6. What is a contour interval?
7. Explain how to determine elevation on a map.
8. Explain how to measure the straight line and curved distance between two points on a map.

Key Words

azimuth
grid azimuth
back azimuth
declination diagram

References

Field Manual 3-25.26, *Map Reading and Land Navigation*. 18 January 2005.

Field Manual 7-8, *Infantry Rifle Platoon and Squad*. Change 1.1 March 2001.

Department of the Army. XVIII Airborne Corps and US Army Center of Military History. (10 June 1991). Operations Desert Shield and Desert Storm. *Oral History Interview DSIT AE 108*. Fort Bragg, NC, and Washington, DC. Retrieved 8 July 2005 from http://www.army.mil/cmh-pg/documents/swa/dsit/DSIT108.htm

INDEX

liberty st.

Matuchen.